CITY

debates on urban sustainability

**Mark Hewitt and
Susannah Hagan**

Published by James & James (Science Publishers) Ltd,
35—37 William Road, London NW1 3ER, UK

© 2001 Mark Hewitt and Susannah Hagan

All rights reserved. No part of this publication may be reproduced, stored in a retrieval system or transmitted in any form or by any means, electronic, mechanical photocopying, recording or otherwise, without the prior written permission of the copyright owner and the publisher.

A catalogue record for this book is available from the British Library.

ISBN 1 902916 17 4

Printed in the UK by Alden Group Limited

Photographs on Cover and Theme title pages: Topoenergetic model of the Inn Valley, created by students of Diploma Unit 6, University of North London. Photographed by David Grandorge.

Contents

Introduction	**VII**

■ Theme 1
City and Culture: overviews of sustainability 1

The city in the post-industrial information society: transforming from a place of production to a place of consumption 2
Peter Weibel

Picture this: conceptual models and sustainable cities 7
Susannah Hagan

From object to procedure 15
Manfred Wolff-Plottegg

Theme 1 Discussion 21

Theme 1 Response: James Caird 29

■ Theme 2
Politics and Planning: winning hearts and minds 37

Mission Impossible: politics and the production of urban space 38
Ian Christie

Car-sharing in Switzerland: a case study 45
Ernst Reinhardt

Theme 2 Discussion 52

Theme 2 Response: Matthew Gandy 57

■ Theme 3
Testing and Modelling: the metabolism of cities 69

The city as natural form: models of urban microclimates 70
Nick Baker

Energy and environmental quality in the urban built environment 80
Matheos Santamouris

Theme 3 Discussion 88

■ Theme 4
Synthesis and Shape: designs on the city 93

Hidden technologies 94
Volker Giencke

Emerging city shapes: energy at the urban scale 100
Mark Hewitt and Andrew Ford

Theme 4 Discussion 109

Theme 4 Response: Katrin Bohn and Andre Viljoen 112

Participants in discussion texts 122

Acknowledgements

The Editors would like to thank the contributors to this volume:

Nick Baker, Katrin Bohn, James Caird, Ian Christie, Andrew Ford, Matthew Gandy, Volker Giencke, Ernst Reinhardt, Matheos Santamouris, Andre Viljoen, Peter Weibel, and Manfred Wolff-Plottegg, and Kester Rattenbury for moderating the symposium.

Mark Hewitt would like to thank:

The students from the University of North London and The Institut für Hochbau und Entwerfen, Innsbruck, (listed opposite) who collaborated on the work for the exhibition "Mountains and Microclimates: Architecture for an Alpine City". The show inspired the symposium "Energy and Urban Strategies", which in turn led to this book. The topoenergetic model of the valley of Innsbruck, which is illustrated in this volume, was conceived, developed and constructed by the students of Diploma Unit 6 with Mark Hewitt and Andy Ford.

Students of Diploma Unit 6, University of North London, 1998-9

Josep Barril
Eva Benito
Lookman Balogun
Stephen Caldeira
Peter Couper
Nicola Dale
Manijeh Dekhordi
Oliver Engley
Luigi Esposito
Ayca Hiz

Namee Im
Mike Menzel
Destine Ozkan
John Pangilinan
Yusuf Quraishi
Paul Saltmarsh
Donald Taylor
Jonathan Tyler
Sam Westman
Mirek Witkowski

Students of the Institut für Hochbau und Entwerfen, Innsbruck

Christoph Gapp
Sabine Gurschler
Thomas Held
David Jenewein
Hubert Lentsch
Reinhard Lins
Henrike Michler
Christoph Milborn

Gerhard Mössmer
Clemens Plank
Carlotta Schmidt-Chiari
Roland Schweiger
Sara Trentini
Nora Vorderwinkler
Michael Wihart

Acknowledgements

The following organisations:

The University of North London have provided major funding for all aspects of the project.

The Institut für Hochbau und Entwerfen, Innsbruck, collaborated on the project, show and symposium.

The Austrian Cultural Institute supported and hosted the symposium and exhibition.

The Vargas Organisation assisted in organising the show and symposium and proposing these events.

Concord Lighting sponsored the topoenergetic model.

The Edge organisation supported the symposium.

Integrated Environmental Solutions for sponsoring the Software for the enviorlab, used in the development of parts of the project.

The following individuals:

Andrew Ford of Fulcrum Consulting who collaborated in the development of the Unit 6 teaching programme at the University of North London.

Professor Volker Giencke, Monica Gogl and Marianne Durig of the Institut für Hochbau und Entwerfen, Innsbruck.

David Ritter and Engineers at Fulcrum Consulting who have contributed to this programme.

Helen Mallinson, former Head of the School of Architecture and Interior Design at the University of North London for constant support and encouragement.

Anthony Auerbach of the Vargas Organisation for helping to develop the show and symposium.

Edward Milford of James and James Publishers who has supported the idea of this "think" book.

And:

Jayne Abel, Robin George, Clare Gerrard, David Grandorge, Chris Hosegood, Gina Mance, Brigitte Marschall, Julie Musgrave, Jane Powell, Joe Scott, Sally Tallant, Ian Whittaker.

The book is dedicated to the students named above, participants at the symposium, and the contributors to this book.

Introduction

This book is the result of a symposium called 'Energy and Urban Strategies', held at the Austrian Cultural Institute on 9 July 1999 with the help of Anthony Auerbach of the Vargas Organisation, and moderated by Kester Rattenbury, writer and journalist. The symposium was the culmination of a year-long collaboration between an Architectual Diploma School design unit run by Mark Hewitt and Andrew Ford at the University of North London, and the Institut für Hochbau und Entwerfen run by Volker Giencke at the University of Innsbruck. The aim of both the design collaboration and the symposium was to develop some sharp ideas about making better cities, in environmental, social and economic terms. The symposium's speakers and guests were from different disciplines and different countries, but with a common interest in the future of the city.

The first theme was entitled 'City and Culture: overviews of sustainability', with three contributors: Peter Weibel, chairman of the ZKM Centre for Art and Media in Karlsruhe; Susannah Hagan, head of a postgraduate programme in sustainability and architecture at the University of East London; and Manfred Wolff-Plottegg, an architect from Graz, with much design experience at the urban scale. Ian Christie, Deputy Director of the think-tank Demos, and Ernst Reinhardt, head of the Transport Section of Energy 2000, Zurich, spoke in the second theme, 'Politics and Planning: winning hearts and minds'. The third theme, 'Testing and Modelling: the metabolism of cities', was addressed by Nick Baker from the Martin Centre for Building Research, Cambridge University, and Matheos Santamouris from the Building Environment Studies Group at the University of Athens.

In the fourth theme, 'Synthesis and Shape: designs on the city', there were contributions from: Volker Giencke from the Institut für Hochbau und Entwerfen, Innsbruck; Andrew Ford, a partner at Fulcrum Consulting; and Mark Hewitt, head of a Diploma unit at the University of North London and partner at d-squared design. In the audience there was a level of expertise that, in some areas, exceeded that of the panel. This was

Introduction

quite deliberate, the intention being to stimulate discussion of a high order.

The symposium was a microcosm of a still-emerging macrocosm, in which nothing is entirely fixed, mapped or agreed upon. This is both the strength and the weakness of 'sustainability' in general and the 'sustainable city' in particular: the looseness of parameters that allows it to embrace an enriching variety of ideas also means it appears to contain too many internal contradictions. Even a cursory reading of the papers will reveal that not even the essentials are yet agreed upon. In a fascinating division along national lines, it was clear that, to the British, the term 'sustainability' meant innovation, change and commitment to the future, and to the Austrians, stasis, conservatism and fear of the future. The British present thus wanted to be identified with the term, while the Austrians wanted to put as much clear blue water as possible between it and them. As a result, the symposium became not so much a dialogue as a duel between two entirely different interpretations of the same word. Each side was surprised by, not to say incredulous of, the other's view.

Interestingly, this did not result in an impasse but in a realisation that while the term 'sustainability' may carry certain irreducible meanings – about revaluing nature, about finding a new co-operative relationship between the built and natural environments, about social as well as environmental justice – those meanings can be interpreted, and implemented – in radically different ways. Sustainability is both regressive and progressive, depending entirely upon who is speaking and/or acting. A Catholic and a Quaker are both Christians; Stalin and Trotsky were both communists; but in each case, the interpretation of the same material was crucially different. The same holds true for sustainability, and for the 'sustainable city'. What became very clear over the course of the afternoon of the symposium was that one cannot refer airily to the 'Sustainable City'. It is not a secular version of the Heavenly City, a Platonic ideal, final, complete and universal, the Answer partially obscured by clouds we are waiting to part. There is no final product, only process, or rather processes, because the interventions necessary for one city to become more environmentally and/or socially sustainable will differ from those necessary for another city, with other historical, physical and social conditions. If and when cities become more environmentally and socially sustainable, they will remain just that – becoming. There is no final stage, no end of history, no perfection. Surely we learned that much from the follies of CIAM and its version of the Heavenly City, eternally new, perfectly efficient?

Physicists like Nick Baker see the city as a complex physical phenomenon, the partial modelling of which will lead to a greater understanding of that complexity, and more energy-efficient designs. Peter Weibel, on the other hand, is a cultural commentator, and views the city through that prism. For him, the physical city is obsolete, rendered irrelevant by information technology. What happens to this 'industrial city' is a matter of indifference, as we must embrace the coming 'information city', which will enable a flow, not only of information, but also of functions. What was once fixed and zoned will now become fluid and mixed. Baker is concerned with having to live with – and in – the rotting carcass of the industrial city in a way that Weibel is not.

Again, however, through the airing of such different points of view, it became clear that each is valid in its own terms: that both the materialist and the culturalist perpectives are 'true', but not the whole truth. Sustainability demands that one think inclusively: that the city, for example, is approached both as a physical phenomenon, and as a cultural construct. The two are inextricably bound up in each other. Cultural attitudes have physical effects. Physical conditions have cultural effects.

The centre/periphery debate throws this relationship into high relief. Ian Christie, for example, privileges the centre, recommending that we try to get those in it to stay and those who have left it to come back. This has unavoidable physical effects – on the centre itself, and on the periphery receiving less attention and resources. Matheos Santamouris, on the other hand, is sceptical of 'recentralisation', and advocates decentralised development. For him, such a cultural decision has beneficial physical consequences: he maintains that less energy is burned in less dense settlements.

Attitudes to the role of the architect in either strategy also vary greatly. There are those like Volker Giencke, an architect himself, who have a traditional view of the architect as form giver, giving shape to the aspirations of a culture. There are others with a more jaundiced, but equally traditional, view of the architect as arrogant, out of touch and unwilling to work with 'the public'. The presence of so many architects at the symposium belied the second view, and the first view, surely, has a few miles left in it, although the process by which these 'forms' are arrived at, and the way they perform environmentally, is of new and central importance.

The audience at the symposium made it very clear that they had more rigorous expectations of the architectural profession than some of the speakers. Their contributions were a vital part of the event, and some of the most interesting and illuminating points were made during the discussion sessions rather than the formal presentations. It is for this reason, probably for the first time in conference proceedings, that some of the discussion has been reproduced here, in addition to the papers, to give the reader a more rounded understanding of what was said. The collisions and confusions are, we believe, a source of encouragement rather than exasperation. What is important is not that the participants failed to arrive at some smooth and polished consensus, but that they plunged with such passion and invention into addressing the vast question of more sustainable cities. Where the discussion focused attention on particular issues that needed elaboration beyond the constraints of available time, we have commissioned 'response' papers. These, in turn, open up some intriguing new questions.

■ **The editors**

Peter Weibel, Susannah Hagan, Manfred Wolff-Plottegg, James Caird

City and Culture
overviews of sustainability

Theme 1

The city in the post-industrial information society: transforming from a place of production to a place of consumption

Peter Weibel
*ZKM Centre for Arts and Media,
Karlsruhe*

Around 1900, the term 'urbanism' was invented to describe a scientific method to control and shape, analyse and study the growth of cities. This method became necessary because the growth of cities had reached a point where they could not be planned any more. They were out of control. The Industrial Revolution generated cities of a size above all prognoses and expectations. Poverty, misery, illness and dirt exploded. Urbanists wanted to find a way to give cities a human face again by planning. The classical conception of a city from which these urbanists started was as a place of production. In the wake of the Industrial Revolution, the city was conceived as a sequence of phases conditioned by industrial labour: you start with an assembly line. Around the assembly line you build a factory. Around the factory you build the homes of the

The city in the post-industrial information society

Peter Weibel

workers. Around the homes you build shops, restaurants and other services.

The publication of a large volume entitled *Une Cité Industrielle* by Tony Garnier in 1907 is a significant milestone in modern town planning. As Le Corbusier said, it was the 'first example of urban land defined as public space and organised to accommodate amenities for the common benefit of the inhabitants (...) integrating housing, work and contact between citizens'. Another title for the project was in fact 'City of Labour', so it is very clear that modern city planning had as a source the idea of labour and production, stemming from the Industrial Revolution. Three main functions were conceived by Garnier: production, housing and health facilities. The dictatorship of production turned housing and health towards the service of production. Workers had to be healthy and housed well to be reliable in the production process. Therefore production is central to the conception of the modern city. Postmodern critiques realised that the city in this sense is not only making the urban environment deteriorate, but is also completely dependent on the environment outside the city. Energy, food, water, nearly everything comes from the non-urban environment. The industrial city is not independent; it does not sustain itself.

Sustainability becomes therefore the central critical argument against the concept of the modern city as a place of production. The 'urban footprint theory' makes it clear that the city leaves many footprints in the environment outside the city – an area 10 to 20 times bigger than the area of the city itself is needed to support it. So it becomes evident that a town focused on industry and production cannot survive. It destroys the surrounding non-urban environment and, in consequence, the city itself. Therefore the post-modern city moved the shopping area to the periphery of the city, and the production and industrial zones outside the city. City centres became vacant: the typical American city became 'bagel city'. The problem of sustainability was only partially solved by the post-modern non-industrial city, however. Responsibility for the supply of water, gas, electricity, information cabling, food, medicine, traffic, sanitary facilities, schools, public services has still to be allotted. Strict reduction of factories and other production sites was a first attempt to reduce the urban footprint in the environment and make the cities sustainable.

At the historic moment when material products of labour lost their pivotal role in the accumulation of capital in the post-industrial society, communication, services and investment became capable of generating more profit than material labour. At this point, the city changed from being a centre of labour to a centre of 'immaterial' labour, like services and communication. Consumption is part of this new kind of urban communication, as shopping is part of consumption. The complete solution to the problems caused by the unsustainability of the modern city came through transforming the city from a place of production into a place of consumption. Post-modern contemporary cities

The city in the post-industrial information society

Peter Weibel

are no longer the places of primary and secondary production, but have become the places of communication, services and transactions. The post-industrial city in the information age has become the knot in a web of universal transmissions, of goods, currencies, messages, information (all kinds of material and immaterial commodities, even cultural commodities).

Consumption in the form of shopping has become a main part of the attraction that cities have today for visitors, and therefore tourists have become more important for cities than inhabitants. The post-modern city learned from Las Vegas: with 40 million tourists a year, it is the post-modern city par excellence. The post-modern city in the rising 'Net society' moves millions of bits of information, which control the supply of food and news, the contents of cultural, administrative, productive and consumptive institutions, in a complex vertical structure, which can only be sustained with the help of computers. Computer sustainability has become the core of the post-modern city. This sustainability is centred on the consumption of cultural or economic goods. The exchange of information, of services, has become the new value, replacing the exchange of products. The exchange of products still exists, but it no longer has the classical function of maximising profit. Profit maximisation can be done today much better in the tertiary sphere of the economy. On this thesis is built the triumph of the new economy over the old economy. Cities of consumption have a new way to regulate the contact between citizens, through services, not labour. If people cannot provide services for each other, they see no reason for communication.

Cultural goods are also subject to these economic laws. Cultural institutions are not measured by the quality of their labour, but by the quantity of their visitors and their profiles for tourist attractions. Cultural institutions turn under the power of economy into institutions of consumption. Event culture, branding, target marketing are not only parts of urban planning but also of cultural institutions. The cities of the new economy are not only temples of consumption, not only paradises of ecstatic shopping for cultural or material goods; they are above all new masks of the market, which make invisible the mechanisms of capital. To tear off the urban masks of capital, we cannot rely any more on culture, because culture has become part of the mask as a privileged way of consumption. We have to go back to the fatal attractors: sin, dirt, hope, chaos. We have to give the cities again the profile of hope. The masses are attracted to cities in the Third World even though they know they will find dirt and poverty. They are attracted by the chaos of the city, because if nothing can be planned, everything is possible. Each individual could be the one out of a million to make a career in the city. Unplanned, uncontrolled cities like those in the Third World, and not the West's city-as-fortress, are the future. The cities of consumption in the post-industrial information age are the new masks of the old fortress.

Picture this: conceptual models and sustainable cities

Susannah Hagan
University of East London

The speeches that accompanied the award of the RIBA Gold Medal to the city of Barcelona, and the recently publicised concerns and recommendations of the Urban Task Force, were quite patently architect-driven: the city is the site of architecture, therefore the city is good and the periphery is bad – culturally and environmentally. This may be so, but people in northern hemisphere cities have been voting with their feet for the periphery for 40 years, even though there are now signs of stabilisation.

Short of making such migration illegal – problematic in democratic societies – one possible solution is to stop thinking of it as necessarily undesirable. To do this, one would have to stop thinking in terms of binary opposites – centre/periphery, architecture/building, brownfield/greenfield, compact/decentralised, good/bad – and adopt a new conceptual model, one that has far-reaching implications for interventions by politicians, planners and architects alike. This model encourages us to think of all types of settlement – urban, suburban, rural – as parts of a continuum or field:

Einstein's unified field theory lends itself usefully as a descriptive analogy ... Here matter which is assumed to inhere ubiquitously in space, identifies itself as fields of relative density or high pressure. Space and object are considered to be made of the same stuff, distinguishable by their relative densities alone ... high pressure denoting objectness, low pressure denoting fieldness ...[1]

This paradigm of greater and lesser intensities, in this case of settlement, could plausibly describe the condition of the city and its periphery, resembling neither the homogeneous dissolution of built culture into nature that characterises Frank Lloyd Wright's Broadacre City, nor the diagrammatic demarcations of Le Corbusier's City of Tomorrow. The 'field' or continuum is not, therefore, an equal and opposite assertion of the virtues of greenfield development over brownfield, or of decentralisation over compaction, but a way of conceiving of the complexity of settlement inclusively, so that sometimes decentralisation might be more appropriate than compaction – even within the city – and sometimes the periphery might be more desirable than the centre. Such inclusivity has obvious implications for the distribution of resources, as all forms of unsustainable settlement would have equal claim to remedy.

Picture this: conceptual models and sustainable cities

Susannah Hagan

To renounce ideas of boundary and privileged difference in favour of this new paradigm of the field is difficult for any of us raised in the Western tradition, and particularly difficult for architects. We have been educated to consider that meaning is created by difference, and the boundary as that which defines the edge between different conditions: inside from outside, city from country, culture from nature. To replace this with a continuum of greater and lesser intensities of settlement, all of which are viewed as equally important, is more than most within architecture can stomach. But it may be advisable if we are to avoid the prevailing crude polarisation between those who look to the centre for a sustainable future and those who look to the periphery.

This is important because neither is going to go away, despite prophecies on both sides to the contrary. For those predicting a cyber-future, life is on the edge, with business and commerce having followed the residential flight to the suburbs to transform them into what Robert Fishman calls 'technoburbs',[2] a 'new kind of decentralised city' reliant upon information technology. These are places of employment and entertainment, consumption and education as well as dwelling, with people often commuting between technoburb and technoburb rather than city and suburb. Spatially and economically, they are 'meshworks', defying prediction and control, and not the hierarchies of the old city centres. The economic health and, it must be said, greater social homogeneity of these technoburbs means that they are not going to wither away just because we turn our fastidious backs on them and address the inner city instead. On the contrary, they will sit there, haemorrhaging energy and pollution, while we address the centre.

Conversely, those who dismiss the traditional centres as redundant are equally wishful, as in many cases these are not withering away either, and for good reason, as Saskia Sassen has demonstrated:

... massive developments in telecommunications and the ascendance of information industries have led analysts and politicians to proclaim the end of cities ... The globalisation of economic activity suggests that place – particularly the type of place represented by cities – no longer matters.

These trends ... represent only half of what is happening. Alongside the well-documented spatial dispersal of economic activities, new forms of territorial centralisation of top-level management and control operations have appeared. National and global markets ... require places where the work of globalisation gets done.[3]

These 'places', according to Sassen's research, are very often the old urban centres their detractors have dismissed as obsolete, and they are found in both the northern and southern hemispheres: London, Paris, New York, Sao Paulo, Buenos Aires, Bangkok, and Mexico City. They have remained what successful cities have always been, 'milieux of innovation', to use Manuel Castells and Peter Hall's[4] term, and, as such, economically if not environmentally sustainable.

What is obsolete, therefore, is to view the centre/periphery debate in terms of 'either/or': privileging either the centre or the periphery when both have strengths and disadvantages, both are highly problematic environmentally, and both are entwined in an embrace that must become mutually beneficial rather than parasitical. For architects, it is difficult to relinquish the privileging of the centre. This reluctance is couched in terms of environmental ethics, but how much of it is in fact cultural prejudice? For example, if the periphery were transformed overnight, if all its buildings used renewable energies, all its rooftops and wasteland were cultivated, all its cars were solar-powered or, better still, abolished in favour of clean and workable public transportation – if, in other words, the democratically desired periphery suddenly became environmentally sustainable as well, many architects would still object to it, on grounds of urbanity, or rather its lack.

An alternative inclusive 'both/and' approach is being preached and practised in various disciplines at various scales. Unlike modernists' prescriptions for the city, the American urbanist Peter Calthorpe does not prescribe. By adopting a radically inductive method, he stands outside the current centrist/decentrist debates, looking at the existing variety of unsustainable dwelling patterns and suggesting ways of retooling them, developing projects in cities, suburbs and new towns that are 'diverse, centered and walkable'.[5] His solutions are based on a pragmatic, case-by-case analysis of different failures for which there is no universally applicable solution. This amelioration takes many forms, depending on pre-existing conditions:

The specific nature of a metropolitan region will dictate how many and which ... growth strategies are necessary and useful. Some regions with a very slow rate of growth may only need incremental infill. Some regions with fast growth and much undeveloped suburban land may benefit from both infill and new growth area projects. Other regions may require all three strategies, including new towns, to absorb massive growth without destroying the identity of existing small towns and urban centres.[6]

At this scale, the social and the environmental are so closely intertwined that Calthorpe would like to see political and physical topographies treated as one: 'At the regional scale, the man-made environment should fit into and along larger natural systems'.

The idea of the social and the environmental merging into one system of demarcation is just one version of the 'blurring' of culture and nature that could benefit both centre and periphery. The object is to ensure that human settlement interferes as little as possible with the ecosystems upon which it imposes itself, so that we cohabit rather than colonise. This is easier to imagine when addressing greenfield development. The datum of a city seems to be a grid or a labyrinth, not soil and subsoil. The connection between the city and the physical given on top of which it perches and within which it sits, is not often made. To make it might start producing interesting new

Picture this: conceptual models and sustainable cities

Susannah Hagan

morphologies of varying intensities, ranging from settlement folded into landscape (the periphery), to landscape folded into settlement (the centre).

In a more specifically architectural context, Bill Dunster's 'Hopetown' is a model for new mixed-use development flexible enough to be useful in urban, suburban or rural development. Hopetown 'proposes a new building type which combines premises for living and working with food production'. What is interesting about this idea is its inclusivity, as evidenced by one model for all occasions. It is an attempt to turn parasitic built fabric into self-sustaining built fabric. By housing our activities in compact forms, Dunster ensures arable land is spared or released, and by terracing roofs so they can be 'intensively gardened', arable land is recreated on top of the footprint of the building:

> Initially the Hopetown model could be used to recolonise urban wasteland … But there is also a need to repopulate the countryside with a compact rural development model…[7]

This 'both/and' approach, in which compaction is as valid a strategy for a rural or suburban context as an urban one, is difficult to implement if one is thinking in terms of binary opposites: centre/dense, periphery/wasteful. The paradigm of the continuum or field helps one to see that what we construct as oppositions are in fact parts of the same ecology of production and consumption. Energy spent fixing the centre is energy only half well spent.

REFERENCES

1 Rhowbotham, K *Field Event/Field Space*, Serial Books, Architecture and Urbanism 2, Black Dog Publishing, London (1999)

2 Fishman, R *Bourgeois Utopias* Basic Books, New York (1987)

3 Sassen, S *Cities in a World Economy* Pine Forge Press, Thousand Oaks, CA (1994)

4 Castells, M and Hall, P 'Technopoles: mines and foundries of the informational technology', in Legates, R T and Stout, F (eds) *The City Reader* Routledge, London (1997) 477

5 Calthorpe, P *The Next American Metropolis* Princeton Architectural Press, Princeton, NJ (1993)

6 Ibid

7 Dunster, W 'Urban sustainability: paradox or possibility?' *AA Files 32* Architectural Association, London (1996)

From object to procedure

Manfred Wolff-Plottegg
Architect

I am an architect who is working with the computer for designing and planning. As I do not like military expressions, I will not talk about strategies. I prefer to talk about algorithms, because in strategies there is a linear one-way subject/object relationship, whereas algorithms provide involvement with relativity.

My centre of interest is not a detached singularity (the aura of an object, the charismatic demi-urge), but a dynamic interchange: architecture and urbanism are public; planning is public – and everything is on the Web. Therefore public and open spaces, which are to be used simultaneously/collectively by several/many people, are the (functional) plot for a merge/morph/scale architecture. Since everything is changing rapidly nowadays, and since I contribute with my work in a 'quick and dirty' way, I basically oppose all strategies of sustainability. When analysing advanced architecture and urbanism, we can see a paradigmatic shift taking place, which is the shift from object to process: instead of planning long-lasting and rigid buildings, we are focusing on evolutionary developments and are, therefore, changing our design methods.

In former times, there have been other architectural rules: for instance, form, function and construction (for 2,000 years!). Later on, there was another rule: form follows function. But the rule of rules is still that a building should fit the architect's (client's) taste. With this rule, architecture is reduced to anthropocentric and expressionistic settings, and because of this rule you see lousy buildings, bad architecture, if you take a look around. So I have been pleading for a long time for a change of rules. As fractal geometry makes clear, it is the algorithm that provides the solution. Just think of the CIAM programme of urbanism and compare it with what Peter Weibel (this volume) has described regarding the development from the functional city to the informational city.

Nowadays we don't talk so much about the rules of architecture; we talk more about design methods. For instance, traditional architecture has been working (and still does) with images/pictures, facades and masterplans. In this field, proportions, materials, colours, aesthetics, 'how buildings look' are important, and it is in this sense that the Etoile was implanted in Paris. But recent developments tell us that we cannot continue by means of 'architectural pictures' – be it a nice-looking facade or a rigid urban grid. We cannot solve the architectural and urban problems of Mexico City, Cairo, Beijing etc. this way.

Nowadays, planning methods are quite different from the measures applied during the 19th century, or even those applied in the middle of the 20th

From object to procedure

Manfred Wolff-Plottegg

century. And we are lucky that they are available, because the problems are quite different. Planning methods have changed radically since we began using the computer in the 1980s, on the basis of the cybernetic architecture of the 1970s. When I talk about using the computer for architectural production, I'm not talking about the use of the computer to imitate what has happened before in planning, or about how architects used to design by hand, by brain or by belly. And I'm not talking about the computer as drafting instrument, or as a tool for image rendering. I am talking about the way in which the computer uses its artificial intelligence, based on new system theories like fractal geometry, chaos theory, fuzzy logic and so on. This is quite different from the ancient methods of architectural design, which has used more or less analogue methods. These new methods are quite probably better able to cope with the current situation.

The central part of the computer is the CPU, the central processing unit, which indicates that the computer is working as a processor: it is processing and controlling processes. To use the computer for processing, as the term CPU indicates, and not to make images, is the right use of it, because rendered images belong, and refer to, the former analogue world of picture-making, when architecture was still a matter of making fixed pictures. Nowadays, we are dealing with computer-processed architecture, because the problems of urban agglomerations are not (and never have been) on the visual/image level, but on the level of controlled/uncontrolled processes.

In this sense, the computer as a paradigm provides us with a new toolbox, and the most important design tasks are to control processes, to simulate processes and to stimulate processes. The most important programs, when using the computer nowadays, are not drawing or rendering ones, but evolutionary models, growth models, game-of-life systems, cellular automata, swarm models, digitally generated architecture etc. There are many different approaches in that direction: 'liquid architecture', 'transarchitecture', 'logorithmic architecture', from John Frazer, Andi Gruber, Chris Langton, Marcos Novak, Lars Spuybroek etc.

The pyramids and the Colosseum, for example, are famous because they survived for a long time. Evidently there is an equation: an architect who wants to become famous must build a famous building, and a famous building must be a long-lasting building. Any tourist will approve of this equation. We all like old whiskies, Rembrandt drawings, traditional habits, and therefore we all like old buildings, and preserve or even reconstruct old buildings. The quality of a building is a function of duration.

These elements come out of a conservative world that has nothing to do with the contemporary evolutionary point of view. If I should write a manifesto for the architecture of the 21st century, I would say that architecture will become non-deterministic, by changing from being the design of objects to the design of procedures. In this digital toolbox with which we are starting a new millennium, I can't find any program that is

From object to procedure
Manfred Wolff-Plottegg

called 'sustainability'. Therefore sustainability is a traditional tool, which is designed to prolong elements which should instead be deleted.

It is a conservative tool that does not create new ideas. It is a tool that wants to keep things running as they are. For instance, everybody is afraid that the globe is becoming overheated, so sustainability is a very good topic for a symposium, and an ideal instrument with which to raise money from Brussels. But architecture and urbanism should be less afraid, and less concerned with sustaining bad existing economic and political systems. They should instead focus on evolutionary models, and other directions where new ideas can be developed. This should prove to be rather simple, because today we know how networking, evolutionary and autocatalytic models are structured, to replace rigid centralised models, and we know how to change from content-related design with aims and desires to the design of systems.

The method of forward planning never looks back; it accelerates architecture. As architecture is a slow medium, which has not kept pace with the major scientific changes of the 20th century, it is even more urgent that new browsing systems become part of our planning methods. To accelerate architecture, design must be made more dynamic. The design process should simulate/stimulate real-life procedures. The forward planning method is not linear, from the first sketch to the 'finished project', from one scale to the next scale, where the finished project differs from the beginning only in terms of details. It is instead modifying continually – not in the sense of

Figure 1 Fictitious detail of an infinitely big number (Plottegg, 1988), demonstrating that there is probably no such thing as sustainability

correcting errors or improving the world, but on behalf of permanent changes of surrounding variables, growth or changing of functions.

An object as building is not the aim, but may be the result. Buildings are changing, and the history of architecture confirms this: after the traditional building-body-architecture (volume, mass), we are nowadays designing information-architecture. Built architecture becomes the flesh for the wireframe-model of reality. The theory of mapping/merging/morphing different electronic, physical and social spaces together as the new 'site' of architecture is the basis for the concept of virtual and real-life space. In a world of self-emerging systems, there are no such things as man-made strategy or sustainability, which are abstruse and old-fashioned concepts.

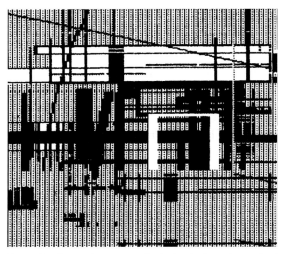

Figure 2 Generative automatic design, 'universal template' (Plottegg, 1987)

Discussion

■ **KATE MACINTOSH** (Architects and Engineers for Social Responsibility) I can hardly believe what I was hearing with this last speaker. Am I to understand that you really think there is nothing to be afraid of in global climate change?

■ **KATE MACINTOSH** You don't think global warming is something we should actually try to avoid since it's mankind that is bringing it about? It is not as though it is happening through some divine intervention; we are doing this.

■ **KATE MACINTOSH** Do you think it is okay? You do realise that if we have the sea rises that are predicted, a very substantial amount of the cities of the world will simply disappear.

■ **MARTIN QUICK** (Architects and Engineers for Social Responsibility) It may well be that the very rich countries can survive climate change by starting again from scratch. But it's usually the poorer people in the poorer countries in the world who suffer from things like climate change. It does seem to me to be absolutely crucial, therefore, that sustainability and energy efficiency are actually at the head of everything to do with cities, suburbs, rural areas; and to dismiss it in

■ **MANFRED WOLFF-PLOTTEGG** You can be afraid. I am more interested in development, and less interested in keeping things as they are.

■ **MANFRED WOLFF-PLOTTEGG** Yes, I think it is happening. Yes.

■ **PETER WEIBEL** Yes. So what?

some theoretical way seems to be really the wrong thing to do.

■ **MANFRED WOLFF-PLOTTEGG** It is not polemical; it is serious.

■ **ANTHONY AUERBACH** (Vargas Organisation) There is a cultural interpretation to what Manfred Plottegg is saying. In a way he is suggesting a kind of cultural dynamic stasis where everything keeps having to revolutionise itself. Is it possible that there is a kind of political/cultural/ ideological thinking that is saying, okay, permanent revolution is over, and so is the kind of permanent life of the rural community; but is there something else? Perhaps there is something in what Peter Weibel was suggesting, that there are new flows in cities.

■ **ANTHONY AUERBACH** Is that possible?

■ **KESTER RATTENBURY** (Moderator) I think there is another issue here, though, which is separate from Manfred's deliberately polemical point.

■ **KESTER RATTENBURY** Does 'sustainability' as a linguistic term imply stasis, and is it then actually a useful term, the word we should be using to describe these things?

■ **PETER WEIBEL** I wasn't talking about permanent revolution. I was talking about evolution. Much simpler, because it is not ideological, you see.

■ **MATTHEW GANDY** (University College London) I think one theme coming through here is that much of the ecological perspective on cities is profoundly conservative, and that's a very important – and ideological – theme to grapple with. Peter Calthorpe, Charles Gence, actually have profoundly conservative views woven into their apparently radical thinking.

Discussion

■ SUSANNAH HAGAN That's the interest of sustainability. It's so synthetic you can be entirely backward looking, static, and want to revive vernacular technology and pre-industrial ways of living, or you can be entirely the opposite, and still be under the rubric of sustainability, because the aim is the same: the aim is so general that the means by which you achieve that can be almost anything. Obviously the people at either pole would sooner die than admit that, because the other pole is totally unacceptable ideologically. To me what's interesting about sustainability is that it can embrace the revolutionary and the conservative, the cutting edge and the traditional.

■ MANFRED WOLFF-PLOTTEGG It is interesting because again it is the idea of having an aim, you see, and what is new and completely different to old-fashioned systems of thinking is that nowadays you can't have a certain aim. Think of the Internet – this doesn't have an aim. There is a big difference in thinking in evolutionary terms and thinking without an aim. Probably there is no aim in evolution.

■ SUSANNAH HAGAN Is there no aim in culture?

■ MANFRED WOLFF-PLOTTEGG In culture, in traditional culture, there has been an aim, and this is the difference now. I feel quite sure that we're now in a different world and we have other ideas. We don't have to follow the various aims which have been existing in former times, and this is a certain freedom which development gives to us.

Theme 1

City and Culture: overviews of sustainability

■ **ROGER BURTON** (Taylor Young Architects)
I think it's interesting in Susannah Hagan's paper that we begin to talk about what is actually happening in suburbia, when the current political thrust is towards looking at the city. The Urban Task Force is talking about the city, how it can reinforce the city, who can move back into the city, how we can achieve a mix in the city. It is actually proving very difficult. We get involved in a lot of regeneration projects which create that mix. Within an urban context it is often difficult. Perhaps we should be saying: allow the city itself to reduce in density. Maybe there is a new suburbanism now. Housing has moved out, and increasingly it's attracting retail and jobs; the schools are there, and we're starting to see clusters emerging almost like sort of stars in the universe, maybe sustainable clusters which actually reduce transportation distances. Perhaps that's the future: allowing the city centres to reduce in density. We should be turning brownfield sites back to green space, increasing the tree planting, trying to reduce urban temperatures, and maybe allow the city to begin to take a back seat. You appeared to express the view that we this is emerging 'naturally'.

■ **SUSANNAH HAGAN** There are definitely technoburbs emerging. Most of the work has been done on what's happening in the States because it's an extreme, but they're not sustainable yet and they're not going to self-organise into sustainable areas. We need to do something about them. There needs to be an aim.

Discussion

■ ROGER BURTON I think we talk about them being technoburbs as if they are dependent on technology; they're not really. It is simply that the facilities are moving out to where the people live. They're not necessarily dependent on technology. We're just living a suburban life which allows people to walk to school, walk to the place of work, walk to the shops. It can actually be very low-tech.

■ SUSANNAH HAGAN That's if it's little businesses like shops, and that's not a technoburb, that's a suburb. Technoburbs are places where corporations have moved to, so that there is a very high level of employment, and they've become – not self-sufficient economically – but serious economic centres. So it's a different animal, if you like, from the suburb, rather more high-powered economically, but just as amorphous. They slide into each other.

■ DAVID CLEWS (University of North London) Isn't the problem of density partly to do with the change in social patterns? The increase in the number of households we need to provide now, the Government's own figure, is 4.4 million. It means that we simply can't provide them as low-density housing. That is what is driving the density – not a view of sustainable cities, simply a numbers game. I think really powerful ideas like this dissolution of the binary [city/country] is a really useful way of thinking about it. If you simply say 'We have to relax the densities', you are back to the problem that you simply can't provide the volume you need.

■ ROGER BURTON I don't think we're saying we're going to relax densities. Some of the suburban development is at increasing densities, wider clusters.

Theme 1

City and Culture: overviews of sustainability

■ SUSANNAH HAGAN The English village is incredibly dense; it's a very efficient form of rural settlement.

■ SUSANNAH HAGAN I mean when people lived *and* worked there.

■ SUSANNAH HAGAN They can't either.

■ PETER WEIBEL Why do you think people go to New York and Bombay and New Mexico, which now has about 20 million inhabitants? The city is built on desire.

■ PETER WEIBEL No, no, no. Work doesn't play a part.

■ KATE MACINTOSH English villages really don't operate as villages any more.

■ KATE MACINTOSH I think there is a huge danger in trying to draw parallels between America and the UK, simply because in America the population-to-land ratio is so much less than it is here, and we simply can't afford those sort of patterns.

■ KATE MACINTOSH They can't either really, and you can't talk about developments without also thinking about transportation. There is a correlation between density of habitation and a viable public transport system. You simply can't get the level of penetration and frequency of service that is needed to make public transport a desirable option in a suburban situation. That is just a fact of life. If you want to find a paradigm for a healthy city, I would say look at Copenhagen or look at Rotterdam. Otherwise we are all going to finish up like Bombay.

■ SUSANNAH HAGAN It's built on work, surely.

■ SUSANNAH HAGAN They aren't moving to Bombay to go to the theatre.

Discussion

■ PETER WEIBEL Not all cities are built around work.

■ PETER WEIBEL No, not true. They are not built on work. These people today don't work, because the money doesn't come from work. We're made to believe that the money comes from labour – but this is too complicated to speak about in architectural circles.

■ PETER WEIBEL It's a dead city, Copenhagen.

■ KATE MACINTOSH What about Barcelona?

■ KATE MACINTOSH No-one is advocating that. We're advocating the Barcelona method.

■ KATE MACINTOSH It's not. It's participation.

■ PETER WEIBEL You have zoning: you have a whole area that has a beach, which is amusement, then you have large public sleeping towns.

■ KATE MACINTOSH And opportunity.

■ MANFRED WOLFF-PLOTTEGG We should accept that there are some developments which we don't condone, like in Mexico. It doesn't mean Copenhagen –

■ MANFRED WOLFF-PLOTTEGG The next point is we cannot pretend to continue planning like, for instance, Haussmann planned Paris. We cannot. These are not tools for city and urban developments nowadays.

■ MANFRED WOLFF-PLOTTEGG You cannot control a city development, an urban process which is running; you cannot control by forms any more, by the shape of Etoile, for example.

■ MANFRED WOLFF-PLOTTEGG It is the same thing.

■ MANFRED WOLFF-PLOTTEGG It's making a great intervention, as has been done historically.

Theme 1

City and Culture: overviews of sustainability

■ KATE MACINTOSH Excuse me, the population density of Barcelona is double that of London. It is a very, very densely inhabited city, right into the centre.

■ KATE MACINTOSH It's very important to get the vitality of the city, commercial variety, richness.

■ KATE MACINTOSH I'm sorry, I don't follow that at all. The secret of the success in Barcelona was the collaboration and the sharing of objectives between the professionals, the politicians, and, eventually, the commercial sector, all facing the same direction for the public benefit.

■ KATE MACINTOSH Property prices in London don't seem to reflect the exodus everybody is talking about. I haven't noticed an exodus. I think the exodus is over.

■ MANFRED WOLFF-PLOTTEGG Higher densities and lower densities – the question of density is a traditional tool of urban regulation.

■ MANFRED WOLFF-PLOTTEGG If you want to raise the average area of a flat, for instance, to 80 square metres, do you want to spread it or make it higher or what? This is not a sociological problem. It is more or less nowadays seen as a mathematical problem. Maybe it could be seen as a transportation problem, but it's more or less a logistical problem, not a problem of a certain social feeling, of having a certain social idea.

■ ROGER BURTON We need density for vitality and vibrancy. We seem to keep hearing that, yet we also hear that people are voting with their feet and going to the suburbs, which implies a lack of vibrancy, lack of vitality. I would love to hear somebody answer that.

Response to Theme 1

James Caird
Head Of Planning for South Shropshire District Council

Introduction

I have been asked to provide a commentary on this first theme, which I do with some trepidation, as I was not present at the symposium and the variety of views expressed was remarkable. In fact, it seems to me that the only possible way of dealing with such a wide range of interrelated issues is to pick out some themes. Before dealing with them, however, I need to stress that my viewpoint is one of an ordinary practitioner living in an ordinary community in England. I acknowledge that opinions on these topics will vary widely but can only comment on the practicalities as I see them.

The character of the debate

I am generally unhappy with the language and terminology that debates on sustainability use. Sustainability will never be achieved without the support of ordinary people worldwide, especially in the West. The arguments will never be won unless ordinary people understand them. It is unhelpful for academics and sustainability specialists to conduct their debates in terms that only they can understand. It is much easier to widen a debate to involve the community if its proponents are already thinking in plain language.

The nature of sustainability

It was noted at the symposium that sustainability meant different things to different people. Four points immediately arise:

1. Unsustainability is not new. If human activity was ever truly sustainable, it stopped being so with the Industrial Revolution and the start of colonialism. It is not credible, therefore, to regard sustainability as some form of balance between man and the rest of nature that can be perpetuated. Sustainability is about lessening unsustainability.

2. The causes of unsustainability are complex and cumulative. The results may be pernicious or catastrophic, and may occur far away from the causes. It is unsafe, therefore, to argue that withdrawal from some activities, which may have unsustainable effects, is synonymous with sustainability. Resorting to vernacular lifestyles, for example, may have some sustainable aspects. But this is not truly sustainable if its proponents rely on support from the rest of the economy, which is unsustainable.

Response to theme 1

James Caird

3 One cannot avoid sustainability issues by redefining the problem. Many types of activity are unsustainable to some degree. Many of the world's resources are finite. Biodiversity is being rapidly eroded. Global warming will give rise to greater and more widespread catastrophes as more places become sensitive to its effects. Localised social and economic pressures will tend to accelerate unsustainable outcomes.

4 It is simply not good enough to adopt the ostrich-like approach of Manfred Wolff-Plottegg. Leaving aside the moral issues, the social repercussions of more widespread detachment of Third World populations from the land that supports them could be catastrophic for the West. It is not just 'their problem'.

Culture and markets

The word 'culture' is unhelpful to the debate. It may reasonably be used to describe the style of a community. But unsustainablility is about people's behaviour, as individuals, which only cumulatively has adverse effects. You cannot express the behaviour of individuals as culture. How they conduct their lives is based on a personal assessment of the alternatives and their costs. This means that sustainability must be about influencing people's behaviour and must, therefore, be about markets. The ability of market pressures to affect sustainability issues will have various constraints:

- People are generally conservative; they will not espouse radical change quickly.

- Markets for new ways of living are, however, ripe for exploitation.

- In a free society, the government is unlikely to have much direct impact on sustainability. It cannot regulate people's freedom of movement and residence. It can merely manipulate the market by fiscal and regulatory means to try to influence the way people behave.

- What is on offer will have to appear to be good value for money as well as offering sustainability.

Location theory

There is no need, in my opinion, to rework classic location theory in this debate. The relative value of pieces of land is dictated by the value of the uses to which they can be put, and to the cost of travel between them. Electronic communication affects the equation only by reducing the need for, or frequency of, some kinds of trip. The idea that this will lead to more diffuse settlement patterns is sound. What is not sound is the idea that this may be more sustainable. A more diffuse settlement hierarchy is almost certain to give rise to more random travel patterns, which will be harder to develop into transportation routes that use energy more efficiently. This is the main reason why the

Response to theme 1

James Caird

thrust of British planning policy has moved towards building upon existing successful and balanced settlements and the main routes between them.

The effects of information technology

The effects of IT on the way we live in future will be profound, but it is wrong to think that IT will somehow supplant normal activities. Few activities consist solely of the transfer of information. Most give rise, in the end, to movements of people or goods. The rate of growth of travel demand is one of the most concerning trends and is the opposite of that predicted for the IT revolution.

Segregation of land uses

Land-use zoning is nothing new. It has existed as long as man has lived in cities. Through most of history, zoning has been a question of practicality rather than design. Land uses differ greatly in their infrastructure requirements and the grain that they impose on the urban fabric. These requirements affect land use distribution today. But by far the greatest effect is on the market for housing. For the past 50 years, people in Britain have not been prepared to accept the housing standards imposed by the urban fabric created by the industrial revolution: workers' housing within a short walk of their work. The population has become middle class and aspires to high standards of living. The suburban fabric we see today is the result of market demand, and not any sort of urban vision. The market will continue to drive land-use patterns. I do not think that this will lead to comprehensive economic specialisms as envisaged by Peter Weibel. While cities have always tended to specialise, there will be no widespread market for places that are so specialist as to exclude functions that are required by the market.

Architects as planners

Architects tend not to make good planners. This is because they tend to see urban problems as opportunities for specific outcomes rather than as the start of a process. Architects' utopias frighten the public, probably because they are clientless. Projects without a client have always had a tendency to get out of hand. There must be someone, with a direct interest in the outcome, to say 'No, I don't want that' or 'That isn't worth what it will cost'. People are worried enough about the impact that economic growth may have on their communities without also having to worry about design-led solutions imposed on them from outside.

Urban design

The importance of urban design cannot be underestimated. By urban design I do not mean the prettying up of swathes of urban fabric to a co-ordinated master plan. Urban design should be

Response to theme 1

James Caird

concerned with freeing up the opportunities to allow the market to develop our cities in a more sustainable way. It concerns town planning to an extent, but is much more to do with economics and politics. It is about creating a market for patterns of urban land use that will be more sustainable and less destructive than the alternatives.

The role of the architect

Architects should forget trying to design society and the distribution of its functions. They should turn their attention very urgently to the design of environmentally friendly and more sustainable buildings. They should, as a profession, be much more proactive about equipping their own members for this task and selling the concepts to the developers, who are generally woefully out of touch with what is required. Developers are afraid of expensive mistakes and must be convinced that the new technologies work and will be cost-effective. Of particular interest are technologies such as material technology, energy efficiency, alternative energy sources, and resource management. The architectural profession must do more to sell these technologies to the public through the media. The public remains fixated on traditional styles and living patterns. More work is needed to sell the new architecture to the 'House and Garden' press. Results will come only from a constant and prolonged campaign to convince the public that more sustainable buildings will work and will save them money.

Theme 1

City and Culture: overviews of sustainability

Summary

I accept the need for architects to be involved in wider sustainability issues, because they are likely to have an impact on everyone everywhere. There is a tendency for the debate to be conducted in ways that are not helpful to the wider understanding of the issues, and even for the issues to be blurred by over-analysis. Architects have a considerable contribution to make in the delivery of more sustainable buildings, but they must not get diverted from this task by issues that neither they, nor anyone else, will have much control over.

Ian Christie, Ernst Reinhardt, Mathew Gandy

Politics and Planning
winning hearts and minds

Theme 2

Mission Impossible: politics and the production of urban space

Ian Christie
Deputy Director, DEMOS

The title of this theme is, I think, politics and planning for the cities of tomorrow, which means the subtitle should be Mission Impossible, because I have got ten minutes to cover what it has taken an urban task force well over a year to do. I am going to talk about the affinities and the tensions between sustainability and the city in political and quality contexts, as linked to the Government's agenda as it is currently emerging around the dreadful vaporous phrases 'joined-up thinking' and 'joined-up policy'. Then I'll look at some of the directions in which we think policy might go in the next ten to twenty years to shape what we at Demos call the 'ecopolis'. This is the buzz-word we're using as a way of summing up a sustainable system of urban economy and urban government.

Mission Impossible: politics and the production of urban space

Ian Christie

I shall be drawing a lot of ideas from *The Richness of Cities*,[1] a series done with Ken Walpole and his colleagues. The other set of ideas I shall be drawing on is a study called *Creating Wealth From Waste*, by Robin Murray.[2] This looks at the ecological, social and economic benefits of waste recycling in cities, and demonstrates that we could at the very minimum get 50,000 jobs out of community recycling schemes.

Let me say something about the political challenge of urban sustainability. I won't try to define it, because there really isn't time, but I think the political equation is reasonably straightforward: it is to find ways of combining pretty drastic reductions in the energy intensity of cities, at the same time as making people want to live in them and enjoy doing so. If you can achieve those simultaneously, then you crack the political problem, and politicians and policy-makers will happily leave the technicalities of achieving sustainability to architects, planners and engineers.

The sustainable city is only just creeping onto the agenda after years in this country in which we have instinctively regarded cities as unsustainable. You have heard some of the reasons why: concentrations of pollution and waste, concentrations of cars – although, in many ways, if you are going to concentrate cars anywhere then the cities are the right place to do it. Energy-hungry, service-intense cities, and also all the socially unsustainable aspects of city life – social fragmentation, social exclusion – often mean the British hate their cities. As somebody who believes in urban sustainability, I have to say that

Theme 2

Politics and Planning: winning hearts and minds

most of the time I also loathe London, even though I know that, to be true to my principles, I really have to stay here. We shouldn't underestimate this loathing. Opinion polls consistently show that if the British could live anywhere, they would rather live in a rural setting, perhaps in a cottage with roses round it. Second choice is a semi-detached or detached house in the suburbs. If they really have to, they'll live in the city, and even then they prefer to live in a house. Living in a flat in the city is absolutely not what most people want.

Obviously there is a huge gap between these people and the policy elite, who read completely misleading documents like the *Guardian* 'Space' supplement, which seems to be written for about 50 people who live in lofts. The assumption is that somehow it is about urbanity, but it is really isn't about sustainable urbanity. It isn't a vision of urbanity that most people would subscribe to; it's completely unreal. Most people hate the versions of city life that we have, including the chic ones, and we shouldn't underestimate that. It adds up to a huge political block in the way of urban sustainability, because even if we arrived at all the technical solutions tomorrow, you would have big problems persuading people that this had anything to do with their own aspirations: where they want to live, where they want to bring their children up. Sustainability is about regulating consumption. We are not very good at that. Liberal democracies like to leave people to make the choices they want to make. A lot of what we have heard requires management in a consensual way, without being completely authoritarian. We have to persuade

Mission Impossible: politics and the production of urban space

Ian Christie

people to manage their own demands, as well come up with solutions which help them do it. That means we're thinking about a very different vision of what many people will do in their careers as architects and planners and engineers.

What is the vision of the sustainable city? Sustainability has numerous attractions for the city. It gives a density that is very attractive, because that density allows us to have diversity without distance. Roger Levett uses the example of a gay Afro-Caribbean cricket lover with a liking for Japanese Noh theatre. This person can have his needs met very well in a city, but very badly in suburban areas and in rural areas. You also get a diversity of services, and can meet an unbelievable number of needs in a very close space. You get efficiencies through resource-sharing in dense communities – at least you can, with good design – and cities are perhaps the best place to use as laboratories for the creation of closed economies. Cities are mines of waste material, something recognised by the waste pickers of Cairo and the dump drivers of Vancouver, and they are also forests of waste paper. So they've got the scale to be centres of recycling, which also gives them the potential to be very efficient centres of learning about how we best link up production and consumption.

There are two other things that don't often get mentioned. First of all, there are many downsides to rural and suburban life. If we start taking sustainable transport policy seriously, and pricing car journeys at an ecologically sensible rate, then suburban living becomes less sustainable and then perhaps less pleasant, giving, over time, a marked

Theme 2

Politics and Planning: winning hearts and minds

incentive to go back to more dense urban living. Second, the political economy of sustainability depends on consumers to a large extent managing their own demands, acting in a mutual fashion, moderating their own desires. That behaviour is most easily talked up, passed on, replicated, and practised in cities. Cities are the classical site of self-restraint, as you rub along with a mass of strangers. People in cities have a mutuality on which urbanity depends, and on which politics and sustainability depend.

At the moment, the vision of the sustainable city means absolutely nothing to the mass of the population, who are still voting with their feet and leaving cities, but the policy tide is beginning to turn in important ways. There is now a call for an urban renaissance, a recognition of the need to limit suburban and rural sprawl, and a recognition of the need to make cities more attractive to people of all ages and all social classes, not just the loft-living fraternity and young people. Those two communities have completely dominated the most recent discussion of the aesthetics of urban living, and pretty unprofitably, I would say. There is a recognition above all, if we are going to win the trust of the public about demand management and attract them back into the cities, of the need to bring them into much more democratic dialogue about design. They have to have a say in their neighbourhoods about what cities should look like, and they have to have a say in the development and reinvention of the planning system. Over the last 30 to 40 years, planning has usually prompted responses like 'How did they let that get built?', or

Mission Impossible: politics and the production of urban space

Ian Christie

'I wouldn't live there if you paid me'. A good test in 20 years' time will be how often we hear that kind of thing if we reinvent planning.

There are some useful indications that should give us a sense of optimism. First of all, the Government is developing many different policy agendas. If it is consistent about bringing them all together, then we might be on the way to something approaching sustainable urban development. Second, a reinvention of the planning system is going on: Local Agenda 21, the modernising of local democracy, is in train. Third, new indicators of sustainability are being developed and taken seriously: the Best Value regime for local authority purchasing, and a new agenda for citizenship education. If we add all those together over a long enough period, I think we get something approaching sustainable urban development.

There are, however, still some key needs that need to be met if we're going to improve the chances of making cities sustainable. First of all, we need to create participation in planning, making dialogue with experts something that is desirable, fun, and valued by citizens. That will only come about, I think, if there is real empowerment of local democracy again, and a real empowerment of new mechanisms within the planning system to make a dialogue between experts and citizens valid and effective. It means rethinking the role of expertise in architecture and planning in a serious way. If I really wanted to be provocative, I'd say that architects have revelled in two roles: first of all, the *auteur*, on the model of a French cinema director or the artist; and second, a hired hand who doesn't have a wider socially responsible perspective on his or her brief. In future I think we need to see architects, landscape designers, planners and civil engineers as subsets of one overall industry that is a sustainable-quality-of-life industry. There should be a common education and training programme for all of those streams, and no-one should come out of it without knowing something of each other's disciplines. At the moment, the gap between planners and architects, and between planners and engineers, is unsustainably wide. We need much closer integration between the public and planners, eco-engineers and policy-makers generally.

If all that works, I think we end up with a vision of the ecopolis where we have a lot more dialogue between citizens, local government, and groups of experts, which experts will have to get used to. We will also, I hope, have a re-industrialisation of cities, with far more labour-intensive jobs coming back and making it more attractive for people to stay in the cities. The key to that is to bring into cities something that only cities can support: highly intensive recycling cities – 50,000 new jobs, more sustainable cities, a nice neat package.

REFERENCES

1   Walpole, K et al. *Richness of Cities Working Papers* DEMOS/Comedia, London (1998/1999)

2   Murray, R *Creating Wealth From Waste* Demos, London (1999)

Car sharing in Switzerland: a case study

Ernst Reinhardt
Head of Transport Section, Energy 2000, Zurich

Modern car sharing was born in Switzerland in 1987, since when it has quickly developed into an ever more customer-friendly and attractive service. As of mid-2000, the world market leader, 'Mobility Car-Sharing Switzerland', will offer decentralised car rental on an hourly basis in over 700 locations in more than 400 communes. Around 38,000 customers will have 1,400 self-service mobility cars available round the clock. Reserve, drive, pay: that's how easy it is. The service is supported by Energy 2000, the Swiss National Energy Efficiency Programme, and its modest financial contribution has helped in product development, market development and business administration since 1992.

Car use declines markedly with this system, resulting in considerable extra demand for public transport services, and in energy savings. The success of car sharing therefore hinges on a good network of public transport and car-sharing stations linked to it, excellent interface

Theme 2

Politics and Planning: winning hearts and minds

management, and reservations systems and national standards. Where these pre-conditions are met, car sharing has great potential.

Research in many cities of Europe has established that more than 90% of the people prefer 'green' modes of transport. Politicians and opinion leaders declare the same preferences. Unfortunately, each group assumes the other to be more car-oriented, which triggers many conflicts. 'Everyone needs, and consequently has access to, auto-mobility' – this seems to be the axiomatic rule in transport planning, but what are the facts?

From recent research done by the Swiss national research programme 'Transportation and Environment', we know that in the cities of Zurich, Bern, and Basle more than 40% of households are carless. 80% of them are happy with this situation, i.e. they have chosen it. How can the interests of this large group of people ever be taken into account by politicians in urban transport planning, when they are not even fully acknowledged by transport planners?

Town planning focuses on car rides of cars that are not driven for the better part of the day, but parked somewhere, taking up valuable public urban space and often hindering modes of green transport. Cars absorb all our attention. Public transport, in contrast, gets much less attention in all the cities I know. A more balanced debate in transport policy would reveal new possibilities: the market share of public transport, bicycles and footpaths taken together has risen from 63% to 75%. The exact reverse is true of the average driving licence holder, who covers three-quarters

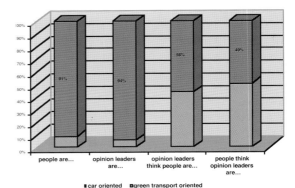

Figure 1 Problems of perception

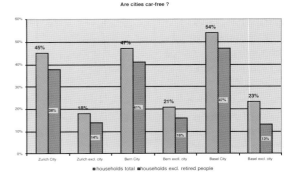

Figure 2 Car-free households in Swiss urban areas

Car sharing in Switzerland

Ernst Reinhardt

of his travelling requirement with individual motor vehicles.

Car-free households substituted borrowing a car from friends with car-sharing, so that, on average, car-sharers used 70% less energy in traffic than other adults.[1]

In 1992 it was estimated that car-sharing in Switzerland represented a sensible alternative to the private car for about 600,000 drivers. Market research in 1997/98 has confirmed this hypothesis (cf. Publications list, Summary). The large individual energy savings and the considerable customer potential make car-sharing an valid target of support for the Energy 2000 Action Programme, which promotes rational energy use and the renewable energies.

Innovative mobility services are provided in close partnership with regional and national transport authorities, so that the public transport system is combined with car-related services for the benefit of the customer. Since 1995, the Zurich Urban Transit Authority (VBZ), in co-operation with the car rental service Europcar, has offered *zri mobil*, a ticket combining public transport and car-sharing. At the national level, 'Mobility Car-Sharing Switzerland' began co-operating with the Swiss Railways (SBB) and the Hertz car rental firm in 1998. *Zri mobil* not only won distinction in the Swiss Design Competition for service design, but also an international award: the first UITP Secretariat General Award for Innovation in Public Transport. Car sharing is proving to be a major agent of change in greening transport. Its appeal to younger people and excellent service orientation make it the first

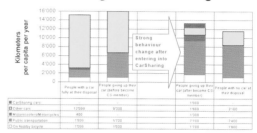

Figure 3 Those who replace car ownership with car sharing enormously reduce their mileage – by 6,700 kilometres or 72% a year.

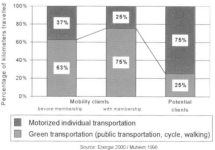

Figure 4 Mobility behaviour changes systematically with car sharing

Theme 2

Politics and Planning: winning hearts and minds

choice for public transport companies in offering seamless transportation services. The Mobility Rail Card 444, which was jointly launched in 1998 by the Swiss Railways (SBB), Energy 2000 and Mobility Car-Sharing Switzerland, opens up for its purchasers, at a price of 444 Swiss francs, access for two years to public transport country-wide at half price. Simultaneously it provides access to all car-sharing cars. Thus, for the first time 'combined mobility' was offered country-wide.

In 1998 it was estimated that 30,000 mobility customers spent 35 million Swiss francs for public transport season tickets – about 3.9 million Swiss francs more than before their entry. Realising this potential would boost the figure to an additional 300 million Swiss francs in earnings from the sale of season tickets.

Theoretically, Switzerland has a market potential for car sharing of 1.7 million people (approximately 24% of the entire population). This includes all those who, based on objective criteria, would benefit financially from the system. To assume that car sharers are poor, immobile or old, and thus could not afford a car, is wrong. The opposite holds true: they are better educated than average; they are richer than average, and they are very mobile in every respect, particularly in their professional lives, engaged in a truly urban life style.

Next to the railway system, the network quality and density of car-sharing outlets make car sharing a key part of the transport infrastructure. We believe in a national scheme, and in national standards irrespective of the size

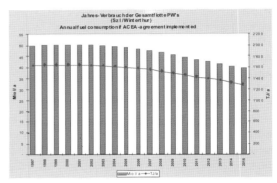

Figure 5 Fuel–saving results from car sharing

Figure 6 Continuous growth of Mobility Car-Sharing Switzerland [5]

Car sharing in Switzerland

Ernst Reinhardt

of the country, since people travel mostly in a regional area. The car is available to everybody all over the globe, and works the same way all over the globe. That's the difference between it and any public transport system. Public transport, after well over 100 years in operation, still needs some expertise to be optimised.

Mobility Car-Sharing Switzerland now serves more than 400 stations, 200 of which are also railway stations. The key to the system, as well as the link to public transportation, is a smart card. It is compatible with EasyRide, the best priced ticket to be introduced from 2004 by the Swiss public transport business.

The characteristics of Swiss car sharing can be summarised as follows: strong customer growth, country-wide coverage, standardised and customer-oriented product range, and simple access to the vehicle fleet by means of the most modern communications technology. For decades the public transport sector has seen little investment. France, with the famous TGV, has paved the way again for a revolution in European railway systems. Above all, however, public transport needs urban, regional and intercity links. It requires a strategy driven by a strong political will to provide a public transportation network.

Distances travelled annually on railway systems reveal stunning differences. In Switzerland, Austria, Germany and the Netherlands, despite having more or less the same culture, they vary substantially. And what might explain the variations between Latin countries? From research we can safely conclude that, amongst other factors, a

Figure 7 Network of Mobility Car-Sharing Switzerland [4]

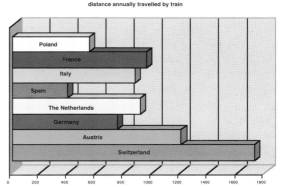

Figure 8 Kilometres per person travelled annually by train, 1995 [7]

dense network with efficient links, system stability and durability, reliability and quality of services are essential to the successful and meaningful greening of transportation. Customer orientation!

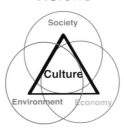

Figure 9 Diagrammatic representation of the relationship between ideas and the market place

REFERENCES

1. Muheim, P and Inderbitzin, J *Das Energiesparpotential des gemeinschaftlichen Gebrauchs von Motorfahrzeugen als Alternative zum Besitz eines eigenen Autos* Eine Untersuchung am Modell der ATG und AutoTeilet Genossenschaft, Lucerne (1992)

2. *CarSharing und der Schüsse; zur kombinierten Mobilität* Swiss Federal Office of Energy, Energy 2000, Bern, September 1998

3. *CarSharing im Urteil von Experten und Opinionleader* Landert Farago Davatz & Partner, Zurich (1998)

4. *Evaluation der Unterstützung des CarSharing durch Energie 2000* IPSO Sozial-, Marketing- und Personalforschung, Dubendorf (1998)

5. *Mobility at Your Convenience/Mobilität wählen/Le choix de la mobilité* Eidg. Dreucksachen- und Materialzentrale (EDMZ), Bern
   Summarizes the main points of the Synthesis in a popular way

6. *Synthese: CarSharing und der Schlüssel zur kombinierten Mobilität* Peter Muheim & Partner, Lucerne (1998)

7   *Summary of Synthesis: CarSharing und der Schlüssel zur kombinierten Mobilität* EDMZ, Bern

8   Reinhardt, E (ed.) *Social Marketing in Transport and Energy Management* Energy 2000/OECD/ECMT, Zürich (1994)

9   *Mobilität in Zürich* Perceptions, Behaviour, Potentials. Socialdata, Zurich (1993) (in German)

10 *Autofreie Haishalte: Ihre Mobilität und die Folgen für die Verkehrsplanung und die Verkehrspolitik* Müller & Romann, Zürich (1999)

11 *Transportation. Yesterday, today, tomorrow. Data, facts, policies. The Swiss transport system and Swiss transport policy* GVF Report 1/98, Swiss Federal Department of the Environment, Transport, Energy and Communications, Berne (1998)

Theme 2

Politics and Planning: winning hearts and minds

Discussion

■ **STEVE JOHNSON** (Architecture Ensemble) [In response to Ian Christie's presentation] I think these discussions have become quite dangerous. We tend to isolate the city from the countryside around. This is particularly dangerous in a country like Britain, which has a higher than average population density and people wanting to live surrounded with rose bushes. In actuality we have got another group of people who want to stay in the city, and I think the real estate values within London reflect that. You would never guess from the cost of housing that people are flooding out of the city. We have got to think of ways in which cities can attract people on a permanent basis. There is nothing more soul-destroying for an architect or an engineer who is designing a housing complex than to know it is going to be occupied during the week and vacated at the weekend. There is so much pressure now being put on rural areas by people living within cities who want to get out of them, decimating the culture of the towns and villages and clearing out inner cities of people who might be supporting those pubs and restaurants, etc.

■ **IAN CHRISTIE** There is a huge mismatch between what people want to achieve for themselves in the short term and the long-term sustainability of their collective choices – for example, moving out to the countryside. If everybody does that, there is no countryside left, but politicians are terrified of saying 'Thou shalt not make this choice' because it looks authoritarian and paternalistic, and it is! There is only one alternative, and that is to find ways in which people can be confronted with the consequences of their choices, and find – in partnership with the planners – ways of reconciling their desires with what is going to be sustainable in the long term. So there really is nothing for it but to give up lots of your evenings in future to inventive debates with local communities. The more we experiment with ways of doing that, the better off we are going to be.

■ IAN CHRISTIE   Yes, it's a complete waste of time as a result.

■ ERNST REINHARDT   To start with, we should differentiate between car sharing and car pooling. The first notion relates to car pooling. Car pooling in Switzerland doesn't work, and I know of no country where it does. Car driving is simply too

■ STEVE JOHNSON   Something like the millennium village at Greenwich has been designed completely in isolation. There has been no discussion.

■ SUSHEEL RAO (Building Research Establishment) A question for Professor Reinhardt: when you mentioned car sharing and looking at who is using green transport, did you include travelling in a car-sharing mechanism as green transport? And is the car sharing actually sharing the cars with other people in the journey, so reducing the number of single-occupancy car journeys, or was it actually people sharing the car, driving off on their own, and bringing it back again? If it's the latter, how would you encourage people to actually share the journeys as well?

fashionable, and car drivers – at least in Switzerland – are very much against critics of the automotive industries. People like driving cars, even if they're stuck in congestion. With car sharing, we have well-established research showing that car drivers drive cars, and car sharers use the transport system: they choose the best option, be it cycling, walking, or sharing a car. Those who use the car-sharing system use about 50% less transport energy. If we build up the whole transport system in Switzerland, it will trigger 300 million Swiss francs right into the pocket of the treasurer of the public transport system.

I might add this: we were talking of a dialogue between government and people, and as a planner I always think we have to establish a dialogue, but I see its limits in terms of time: getting people's attention, keeping them at the table, arriving at a result, something down to earth – all this takes time. Nor should we forget public/private partnerships. We in the publicly run programme, Energy 2000, started off by talking to politicians and associations of politicians and lobbies. Now we work almost exclusively with private companies, and we can move at a faster pace. We can reach people, not through public debate, but through services and goods for which they are willing to pay.

■ KESTER RATTENBURY Yes, I certainly agree with that. My impression of participation, from my slight involvement, is that it's quite a crude tool, and it is certainly only valuable if the support system is there and if there is a meaningful kind of chance that it's going to be used. It can't simply be done for its own sake.

Discussion

■ KATE MACINTOSH I would like to take issue with Ian Christie about the British love affair with suburbia. I think that where people want to live is all about access and choices, and if people are voting with their feet to leave the big cities, it's largely because of quality of education and better quality of other environments. The area of London we're meeting in at the moment, insofar as it's a residential area, is largely flats. There is no difficulty with uptake on the accommodation here. Likewise Scotland. Everyone will say that Scottish culture is different, but people like to live, want to live, in the heart of Edinburgh, in the heart of Glasgow: very dense urban living, because it's high quality in every way. They have access to public spaces, quasi-public spaces, even though they're in a flat.

■ BRIAN MARK (Fulcrum Consulting) There has been a lot of discussion about empowerment and inclusion and demands for control, but I think there is a huge danger that this is just lip service. There is no education programme to let the general public know about the issues; there is virtually nothing in the national curriculum or on television. There are hours and hours of rag-rolling your living-room, to the extent that it has actually changed the whole buying pattern of DIY

■ IAN CHRISTIE You are absolutely right; there is a close connection between the quality of education and choices about moving out, and we're currently doing a project looking at how you can link education and housing policy in such a way that you can attract people back into cities to form more mixed neighbourhoods and have much more socially mixed schools.

Theme 2

Politics and Planning: winning hearts and minds

products, but it is useless for educating the general public about environmental issues. How can anyone become involved if they don't know about the issues?

■ **IAN CHRISTIE** One thing we can do is influence how government designs its new programme for citizen education to be introduced in 2002. That gives us all two and a half years to make sure it doesn't become a bolt-on civics lesson, which would be a complete disaster.

Response to Theme 2

Matthew Gandy
University College London

Figure 1 Refuse collection service in Beckett Street, Borough in 1903. Waste collection formed part of a wider crusade for improved urban sanitation and until recently was generally carried out by municipal government. Source: Greater London History Library.

> The system works only if waste is produced... Post-industrialism recycles; therefore it needs its waste.
>
> Giuliana Bruno[1]

> Recycling, regardless of its limitations, is acclaimed as an emerging alternative, that will soon reduce the need for landfills. Between the threat of the present and the promise of the future, the practice of the past is reproduced.
>
> Stephen Horton[2]

The last 50 years have seen a dramatic increase in the municipal waste stream of Western economies. This can be attributed to a number of factors: rising levels of affluence, cheaper and less durable products, the proliferation of packaging, changing patterns of taste and consumption, and growing demands for convenience products. This 'waste mountain' has led to a new emphasis on organic conceptions of cities derived from 19th-century anxieties over the loss of valuable materials. Whereas the motivation for recycling activities in the past was overwhelmingly

Figure 2 Hand picking of refuse at Cookson's Dust Yard, Tinworth Street, Vauxhall in 1904. In the early twentieth century many people made their living from sorting wastes in Europe and North America. Source: Greater London History Library.

economic, the current emphasis has sought to combine a range of social, economic and environmental rationales.[3] For Ian Christie of the London-based DEMOS thinktank, recycling now holds the key not only to environmental sustainability but also to urban regeneration: the ecological and the economic have been fused to produce a radical policy agenda for 'highly intensive recycling cities.'[4] In this paper I shall argue that the current emphasis on recycling is misplaced and serves to divert our attention from a more critically engaged analysis of urban environmental issues.

Since the 1960s the reduction of waste through the promotion of recycling has steadily worked its way up the environmental policy agenda to form a cornerstone of most contemporary perspectives on urban sustainability. Yet these organic or ecological models of urbanisation harbour a series of practical and political contradictions that are overlooked in most of the existing environmental literature. Many environmentalists assume that there will be an inevitable shift from our 'throwaway' society to a post-industrial 'recycling' society of the future.[4] This sentiment can be illustrated by the recent DEMOS study by Robin Murray entitled *Creating Wealth from Waste*: 'Matter is now assuming an equivalent status to time, and applying intelligence to the way we treat and use materials is the great challenge of the next industrial revolution'.[5]

I want to look at the idea of materials recycling in greater detail as part of a wider critique of what we might term 'ecological urbanism'. The current

Response to theme 2

Matthew Gandy

Figure 3 Collection vehicle for paper and waste food operated by the Metropolitan Borough of Poplar in 1954. Elaborate waste collection schemes were at their peak in war time but, for mainly economic reasons, gradually faded out in the 1960s and 1970s. Source: Greater London History Library.

Figure 4 Collection facility for waste food operated by the Metropolitan Borough of Poplar in 1954. Food wastes were commonly used for pig feed until, in the 1960s, this practice was proved to cause the spread of pathogens to humans. Source: Greater London History Library.

emphasis on recycling needs to be placed in the context of two interrelated developments affecting the scope of environmental policy-making. A first issue is the emphasis on individual action for environmental protection rather than government or state regulation in the promotion of environmental protection. The second related issue is the growing role of the private sector in municipal waste management. The privatisation of waste management can be traced to the mid-1970s, when the existing structure of municipal waste management began to be increasingly criticised on a number of grounds: the growing environmental controls and technical complexity that left many local municipalities only collecting waste; the spiralling costs of waste collection and waste disposal, which were a growing tax burden; and the complexity of new developments in waste management, which was exceeding available expertise in local government.[6]

The capacity of the private sector to take on operational aspects of waste management has developed substantially since the 1960s and 1970s in contrast with the ramshackle efforts of the past, which played a key role in fostering the emergence of the public health movements and wider calls for the reform of local government. Large firms, including multinational companies specialising in the provision of municipal services, are able to benefit from economies of scale and undertake research into profitable aspects of waste management such as large-scale landfill and incineration technologies increasingly beyond the operational scope and financial capacity of local

Theme 2

Politics and Planning: winning hearts and minds

Figure 5 The Edmonton incineration plant in Enfield, North London. 1992. This plant, which has been operated since the early 1970s, has recently been the focus of political concerns over air pollution. Source: Matthew Gandy.

government. A profit-driven global consolidation of the waste/water/environmental technology sectors reflects the vast new markets that have been created by successive waves of environmental protection legislation since the 1980s.[7]

The dominant methods of waste disposal for household waste remain landfill and incineration, yet proponents of recycling suggest that these options can eventually be displaced. The continuing reliance on landfill has been increasingly criticised as a viable disposal option because it produces toxic leachates that can contaminate water supplies, and produces combustible landfill gas from the anaerobic decomposition of waste.[8] Consequently, the new generation of landfill sites are larger, and further away from centres of population; they are much more closely monitored, and their operation often involves advanced technical methods of pollution control. As for incineration, its widespread use began to decline sharply from the 1920s onwards because of competition from cheaper waste disposal by landfill. In addition to the economic disadvantages of incineration, these facilities have been widely criticised since the 1960s for their impact on urban air pollution. In the 1980s and early 1990s there was a resurgence of interest because of a combination of improved design, enhanced levels of profitability and a political consensus in favour of the expansion of non-fossil-fuel sources of energy. Yet this 'incineration renaissance' has been cut short by new evidence of the health risks from dioxins produced from the burning of chlorine-containing compounds such as plastics and bleached paper.[9]

Response to theme 2

Matthew Gandy

Figure 6 Hamburg, 1990. Consumption and waste: an intractable dynamic? Source: Matthew Gandy.

The widely held contention that landfill is a marginalised, unprofitable and anachronistic dimension to waste management is misleading. The closely monitored and often purpose-built landfills of today owe little to the rubbish tips of the past. A new geography of pollution is now being driven by land use, property prices and inequalities in political power. In the United States, for example, we find that rural minority-dominated backwaters in Virginia and Pennsylvania are the new repositories for urban wastes for which no locally acceptable political solutions can be found. More and more wastes are moved from affluent municipalities to impoverished communities who have little to trade but cheap land in return for new sources of income and lower taxes.[10] The frequent analogy of cities as 'urban mines' suggests that the loop between production and consumption can be closed within each metropolitan arena. Yet cities are nodal points within a wider set of flows and processes. With the increasing globalisation of trade, for example, regional returnable systems for the recycling of packaging become progressively more difficult to operate effectively. The rapid colonisation of Eastern European packaging markets provides an illustration of this process as long-established recycling systems were swept aside during the 1990s by new forms of packaging such as laminated cartons.[11]

An important economic benefit claimed for recycling is that it may reduce the costs of waste disposal for urban areas that have fewer cheap landfill opportunities. Yet the costs of

Theme 2

Politics and Planning: winning hearts and minds

Figure 7 Hamburg, 1990. Should the modern city aspire to the organic urbanism of the past? Source: Matthew Gandy.

comprehensive materials recycling programmes have been consistently underestimated in relation to landfill and incineration. A consideration of the economic aspects of recycling is important because a key reason for the promotion of recycling is frequently argued to be the potential of schemes to generate a profit from the sales of materials and the creation of savings in waste disposal costs. The recycling of some materials such as glass via the 'bring' system of bottle banks may be economically viable in some cases, but most evidence suggests that comprehensive recycling programmes where kerbside 'collect' systems are extended to paper, putrescibles, plastics and other materials are more expensive than routine waste collection for disposal by landfill or incineration. Recycling is a service that must vie for public money with other priorities such as education or housing.[12]

In order to understand the limitations of post-consumer recycling it is useful to picture recycling as a hierarchy of potential options ranging from waste reduction at source to energy recovery from landfill or incineration. The emphasis on waste prevention in the production process is integral to radical environmentalist demands to challenge the underlying economic dynamics for the creation of waste. For example, the Berlin-based Institute for Ecological Recycling claims that up to 90% of pollution results from the production process, and not in the eventual disposal of products, and this underlies the rationale for a focus on waste reduction at source rather than a policy emphasis on materials in post-consumer waste.[13] The emphasis on waste reduction is

Response to theme 2

Matthew Gandy

critical because most recycling debate revolves around different proportions of a growing waste stream. The DEMOS study, for example, is overwhelmingly focused on percentage targets for recycling rather than any wider examination of the long-term dynamics for waste production.[14]

A further claim for recycling is that it has the potential to contribute towards employment creation and the economic regeneration of cities.[15] Yet the vast majority of these jobs created would be low skill and low pay: to suggest otherwise reveals a fundamental contradiction between the pay and conditions of a 'recycling economy' and the wider economic rationale for the expansion of recycling in the face of cheaper alternatives. The urban recycling economy would promise little more than a form of 'ecological workfare' for those excluded from alternative opportunities in the labour market. Few middle-class environmentalists can be expected to choose a poorly paid career in waste picking. The return of nineteenth-century 'scow trimmers' or the emulation of the scavenging communities of contemporary Third World dumps presents a disturbing model for social cohesion in the twenty-first century. Indeed, the structural dimensions to urban poverty are scarcely addressed in much of the urban sustainability literature.

In addition to the high cost of multi-material recycling programmes, the other main economic barrier is derived from the weakness and volatility of the secondary materials market, which has adversely affected the scope of recycling since the 19th century. In 1895, for example, the American

Figure 8 Multi-material recycling centre, Hamburg, 1990. These environmental services do not save money but are provided to meet public demand. Source: Matthew Gandy.

Theme 2

Politics and Planning: winning hearts and minds

Figure 9 Utopia or eyesore? Bottle banks in Islington, London 1992. People who take their waste glass to bottle banks by car may use up more energy than is saved through recycling. Source: Matthew Gandy.

sanitation pioneer George E. Waring lamented, 'We have not yet reached any very satisfactory knowledge as to the conversion of waste into wealth. While the theoretical value of discarded matters is recognized, the cost of recovery is still an obstacle to its profitable development'[16]. The weakness of secondary materials markets continues to affect recycling in three main ways: firstly, the periodic collapse of markets in response to increases in the amount of collected materials during periods of environmental concern; secondly, the long-term decline in some sectors, such as rubber and textiles, in response to changing patterns of production and consumption and advances in raw materials processing technology; and, finally, the innate weakness of the secondary materials market, reflected in the impact of economic recession on the demand for waste paper and other commodities. These difficulties are scarcely addressed by most programmes to radically increase the level of recycling towards technically achievable levels, and the problem is merely compounded by more effective collection systems in the absence of an intricately co-ordinated and planned attempt to influence the whole cycle of production and consumption as occurred with the introduction of the German Grüne Punkt system in the early 1990s.[17]

The history of post-war recycling initiatives can be read as a gradual transfer of waste collection responsibilities away from local government towards in the first instance voluntary groups and various types of shared collection facilities in the 1970s, followed by the extensive emphasis on

Response to theme 2

Matthew Gandy

Figure 10 High rise housing, Lambeth, London 1999. The real challenge for urban recycling is to extend collection schemes into high rise housing developments and promote shared composting facilities. These services are technically feasible but economically expensive. Source: Matthew Gandy.

individual consumers in the 1980s as part of the wider rationale for 'green consumerism'. By the 1990s, however, it became clear that earlier approaches had failed to make a significant impact, necessitating a fundamental rethink of the mechanisms by which the scale of recycling might be increased without incurring higher levels of public spending. The solution that has emerged in Europe has been based on forcing the producers of packaging waste to collect and recycle their own products through private organisations. In reality this emphasis on 'producer responsibility' has simply forced the costs of these more elaborate collection systems back onto consumers whilst at the same time raising the spectre of a vast and inadequately regulated trade in recyclable materials as waste carriers seek to illegally dispose of material for which no viable market can be found.

We must conclude that the prospects for far higher levels of recycling are not good. The recycling of household waste does not save money and is of only marginal environmental significance. The emphasis on recycling perpetuates 19th-century urban metaphors in the service of twenty-first century global capitalism. A convincing alternative to current waste management strategies will have to fully incorporate wider issues surrounding the geography of waste disposal and the restructuring of the urban labour market. If urban sustainability is to link ecological concerns with those of social justice then purely cosmetic responses to the future of urban waste management will have to be rejected.

REFERENCES

1 Bruno, G 'Ramble city: postmodernism and *Blade Runner*' 1987 October 41, 61

2 Horton, S 'Rethinking recycling: the politics of the waste crisis' *Capitalism, Nature, Socialism* 1995 6 1–19

3 See: Gandy, M *Recycling and the Politics of Urban Waste* Earthscan, London/St Martin's Press, New York (1994); Galbraith, J K *The Affluent Society* Penguin, Harmondsworth (1958); Packard, V *The Wastemakers* David McKay, London (1960)

4 Examples include: Young, J E *Discarding the Throwaway Society* Worldwatch Institute, Washington, DC (1991); Kharbanda, O P and Stallworthy, E A *Waste Management: Towards a Sustainable Society* Greenwood Press, Westport, CT (1990); Brown, L R and Jacobson, J L *The Future of Urbanization: Facing the Ecological and Economic Constraints* Worldwatch Institute, Washington, DC (1987); Platt, B *Beyond 40%: Record-Setting Recycling and Composting Programs* Institute for Self-Reliance/Island Press, Washington (1990); Pollock, C *Mining Urban Wastes: The Potential for Recycling* Worldwatch Paper 76, Worldwatch Institute, Washington, DC (1987); Elkin, T and McLaren, D *Reviving the City: Towards Sustainable Urban Development* Friends of the Earth/Policy Studies Institute, London (1991)

5 Murray, R *Creating Wealth from Waste* DEMOS, London (1999) p 35

6 See, for example: Goddard, H C *Managing Solid Wastes: Economics, Technology and Institutions* Praeger, New York (1975); Savas, E S 'Policy analysis for local government: public versus private refuse collection?'

Response to theme 2

Matthew Gandy

Policy Analysis 1977 3 1–26; Savas, E S 'Public versus private refuse collection: a critical review of the evidence' *Urban Analysis* 1979 6 1–13

7 Harvey, D *Justice, Nature and the Geography of Difference* Basil Blackwell, Oxford (1996)

8 Tucker, D G 'Refuse destructors and their use for generating electricity: a century of development' *Industrial Archeology Review* 1977 1 74–96. See also: Melosi, M V *Garbage in the Cities: Refuse, Reform, and the Environment, 1880-1980* Texas A&M University Press (1981)

9 See, for example: Hoffman, R E 'Health effects of long-term exposure to 2,3,7,8 tetrachlorodibenzo-p-dioxin' *Journal of the American Medical Association* 1986 April 460–493; Spill, E and Wingert, E (eds) *Brennpunkt Müll* Sternbuch, Hamburg (1990); Rappe, C, Choudary, G and Keith, L H (eds) *Chlorinated Dioxins and Dibenzofurans in Perspective* Lewis Publishers, Chelsea, MI (1986)

10 Gandy, M *Concrete and Clay: Reworking Nature in New York City* MIT Press, Cambridge, MA (in press)

11 Heering, H, van Geldner, J W and Blaauw, J A *Tetra Pak in Eastern Europe* Friends of the Earth, Amsterdam (1992)

12 Gandy, *Concrete and Clay*

13 Institut für Ökologisches Recycling *Abfall Vermeiden* Fischer Taschenbuch Verlag, Frankfurt am Main (1988); Institut für Ökologisches Recycling *Ökologische Abfallwirtschaft: Umweltvorsage durch Abfallvermeidung* IföR, Berlin (1989)

14 Murray *Wealth from Waste*

15 Ibid. See also: Vogler, J *Work from Waste: Recycling Wastes to Create Employment* Oxfam/Intermediate Technology Publications, London (1981)

16 Waring, G E 'The disposal of a city's waste' *North American Review* 1895 July 49

17 See, for example: Greenpeace Toxic Trade Campaign *Plastic Waste to Indonesia: The Invasion of Little Green Dots* Greenpeace, Hamburg (1993). See also the general critique of Western recycling by Luke, T J 'Green consumerism: ecology and the rise of recycling' in Bennett, J and Chaloupka, W (eds) *In the Nature of Things: Language, Politics, and the Environment* University of Minnesota Press, Minneapolis (1993) pp 154–172

Nick Baker, Matheos Santamouris

Testing and Modelling
the metabolism of cities

Theme 3

The city as natural form: Models of urban microclimates

Nick Baker
*Martin Centre for Building Research,
University of Cambridge*

Preamble about sustainability

To start on a provocative note, I feel that the notion of the sustainability of a city is a contradiction in terms: cities are concentrations where there is a flow of resources in and out, so no city can be self-sustaining. However, a city as a part of a larger sustainable system is a more realistic concept. But I would also question whether we want things to be sustainable. We could have sustainable traffic jams in London, thousands of cars, buses and lorries gridlocked but using nice clean solar-generated hydrogen fuel! We could imagine these sustainable vehicles scorching up the motorways, their owners occupying totally dispersed and unrelated regions for work, leisure and residence.

The city as natural form

Nick Baker

What else would happen if energy were free and clean? Would our buildings and cities automatically be wonderful places? If artificial lighting and air-conditioning had neither financial nor environmental cost, would we not see a proliferation of the megabuilding and the megacity, where humans become increasingly isolated from the natural world outside, the environment that shaped our genes? We could have sustainable poverty and starvation – indeed we seem to be achieving this already. Few would advocate this as an objective, although it could logically figure in a purely 'technical' strategy for sustainability.

My point is that the concept of sustainability is a diffuse one, lying somewhere between the strict technical meaning of the word 'to maintain or prolong' and a kind of moral catch-all, meaning 'things that we think are good for us and the environment ... probably'. So I think 'sustainability' is a temporary word, and we must not underestimate the importance of the definition of words. Big words that mean different things to different people are one of the most dangerous devices man has invented, as the history of politics and religion clearly demonstrate. Already, improbable things are being done in the name of sustainability.

The ranting against 'the word', however, only serves to underline the fact that we have a very incomplete understanding of the dynamics of human habitation on this earth, mainly because we see it from a single viewpoint in cultural space and time. The city is such a strong case in point – so overlaid with cultural, economic and

Theme 3

Testing and Modelling: the metabolism of cities

political meanings that we rarely stand back and look at how humans occupy the physical matrix of the city. Perhaps we should spend a little time looking at the city like an entomologist looks at an ant-hill. He can't interview the ants, or consult their newspapers, or visit their museums, but by careful analysis of their circulation, the inflow and outflow of material, and the microclimate, he can find out a lot about their necessities for life.

Energy and the city

We have probably all seen those amazing satellite pictures of the surface of the globe at night (Fig. 1). A visiting alien would probably jump to the conclusion that here were massive concentrations of human life, burning off huge quantities of energy. Although this is a manifestation of just one aspect, artificial lighting, the conclusion would be broadly correct – but is this due simply to the concentration of population, and how does this effect the energy consumption per capita? It is easy to find diametrically opposite views. For example, on the one hand we hear that the compact city offers huge benefits in reducing transport energy (assuming that people live close to their work), but, on the other hand, it is the dense cities that most people try to avoid for the location of their homes, creating huge increases in transport demand. There are parallel arguments for and against in the urban built environment. Do urban buildings consume more or less energy, and can the urban microclimate be beneficial to both

Figure 1 Europe at night from the sky. A composite image from satellite photographs.
Source: http://www.phy.mtu.edu/apod/ap990516.html

energy use and quality of life, if the key parameters are right?

Urban analysis at the Martin Centre

To some extent this is the approach at the Martin Centre in Cambridge, which dates back several decades to the work of March and Steadman.[1] In this, simplified generic urban forms were identified and their environmental performance assessed. In our current research,[2] modern image-processing techniques have enabled us to analyse real urban forms, since we now feel that it is the mixture of order and disorder that may be a key factor in environmental quality. Translating from generalised forms to real urban tissue has always been a problem. Furthermore we have extended the area of interest beyond just the energy use of the building, to the environmental quality of the spaces between the buildings. Thus microclimatic issues such as wind and pollution dispersal, penetration of solar energy into the streets, and the impact of open green spaces, are now included in our studies.

Urban texture and the urban porosity model

Aerial pictures of cities show an extraordinary geometry, with striking differences in texture (Fig. 2), a mixture of regularity – right angles and vertical surfaces – and randomness. If we were viewing this as a biologist views the habitat of a

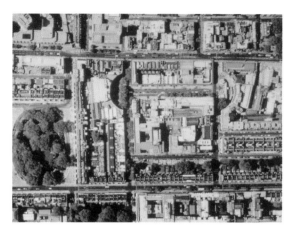

Figure 2 Victorian London (above) and Haussmann's Paris (below), showing striking differences in geometry

Theme 3

Testing and Modelling: the metabolism of cities

newly discovered species, we would find a significance in this geometry. Is there, then, something inherently environmentally efficient – or not – about certain urban textures?

Before proceeding at the urban scale, let us consider Figure 3, which shows the energy use of 92 office buildings in Britain.[3] These offices are full of people looking at VDUs and pushing pieces of paper around, and yet there is an enormous (twenty-fold) variation in energy use. What is the cause (or causes) of this wide variation? Is it possibly due to the location of the building in a block? Or to what is opposite? Or to whether it's near a green space, or in a very crowded area? Are there significant direct effects, such as the urban heat island effect increasing the need for air-conditioning, or the overshadowing of other buildings increasing the need for artificial lighting? Or are there secondary effects that involve behavioural issues: street noise and pollution affecting the potential for natural ventilation, which in turn affects energy use? The quality of outdoor space also affects energy use: people are more likely to wait for a bus if it's pleasant, than they are if they're standing in an overheated polluted and noisy street.

Our energy modelling studies suggest that, in order to account for the wide variation of energy use, there are many subtle interactions and non-linear effects, some of which may be dependent on urban texture. We have only made a start at investigating these complex effects, and I shall now describe a few of these.

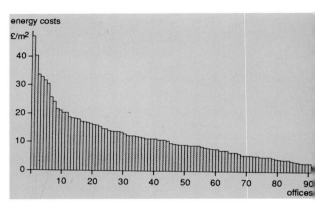

Figure 3 Annual energy costs in 92 individual office buildings. (Energy cost relates closely to primary energy use and CO_2 production). Source: BRECSU.

The city as natural form
Nick Baker

Natural ventilation

One of our current European projects is PRECiS (Potential for Renewable Energy in Cities).[4] One of the five participants has been doing some computational fluid dynamics (CFD) analysis of the pressure coefficient across faces of buildings, for a new development area in Trondheim, Norway. The pressure coefficient indicates the magnitude of the wind-generated pressure on the building. However, we are not looking at just the average magnitude of it; what matters is how much it varies, both in time and space, over the surfaces of the buildings. Figure 4 shows the distribution of pressure coefficients in false colour, over the surfaces of a simplified array of buildings. This could, of course, be a real part of a city, and it could give us an indication of the potential for natural ventilation to avoid air-conditioning as air-conditioned buildings routinely consume twice or three times as much as non-air-conditioned buildings.

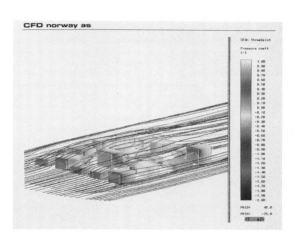

Figure 4 Results of mapping (wind) pressure coefficient on the surfaces of buildings to indicate potential for natural ventilation. PRECiS study site in Trondheim. Source CFD Norway.

Heat island effect

Another thing that we have been investigating is the heating up of the urban environment, the so-called 'heat island effect'. This is primarily due to the absorption of solar radiation, and we have been investigating the impact of the reflectance and texture of the city on this process. Figure 5 is part of a model of an area near Tottenham Court Road in London, which has been painted different colours to see whether the absorption of

Figure 5 Physical model of London study site being tested for optical reflectance in heat island study.

radiation at different angles is significantly affected by reflectivity and urban geometry. Comparing this with an area in Berlin and an area in Toulouse, we found that the absorption varied by about 30%. This could lead to a variation in urban temperature of about 1.5 °C, which in turn makes a significant impact on air-conditioning loads and hence energy use.

Image processing techniques

In order to describe the vertical dimension of the city plan, we have been using a technique where the two-dimensional map is given a third dimension — something called a digital elevation model (DEM). When displayed, we see the height of the building in shades of grey, so the tallest buildings are black. Using image processing software[5] initially developed to enhance photographs, we can carry out a wide number of geometric analyses. One of these we call the 'directional permeability'. Referring to Figure 6, imagine a sectional plane (the green line in plan) moving through the city. As it cuts into taller buildings and then lower buildings and streets, the mean height of the urban fabric along the section oscillates as in the lower graph. If you swing the green line around so that its orientation coincides with the street directions, the mean height would oscillate with a much greater amplitude as in the upper graph. From the amplitude of these oscillations, we can automatically determine the directions of greatest permeability. This can be carried out from any

Permeability

London

N - S sections average height

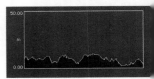

NW - SE sections average height

Figure 6 Digital elevation model (DEM) of London study site with mean height profiles for two different directions, illustrating derivation of permeability rose.

The city as natural form

Nick Baker

digital elevation model by a macro written in the image processing software.

Figure 7 shows our three study areas: London, Toulouse and Berlin, chosen for their apparently different textures. It also shows the 'permeability roses' that have been generated by this method, and which clearly suggest that permeability may be significant for the clearing of pollution by the prevailing wind of the region, and may also have an impact on the trapping of solar radiation. The image processing technique can also be used to map hypothetical building energy use. In Figure 8 the false colour represents the minimum primary energy use of the buildings. The impact of overshadowing of direct sun and daylight by adjacent buildings can easily be recognised. Though the variation in energy use is quite interesting, comparison between our three case study sites shows that there is only a difference of about 13% in the average energy consumption.

Conclusion

The pace of global warming continues to speed up. It does not pause to give us time for our final answer on sustainability. It is almost universally agreed that the use of fossil energy is the main cause. Conventional energy analysis simplifies buildings' context and operation. New computational techniques begin to address the complexity of real context, and thus may begin to explain the wide variance of energy use in real buildings, though it remains to be seen whether

Figure 7 Permeability roses for three sites in London, Toulouse and Berlin.

Theme 3

Testing and Modelling: the metabolism of cities

these techniques will yield answers to key questions about the urban environment.

Returning to the question of sustainability, it is far from clear what the key questions are. However, the negative aspects of the urban environment are generally uncontentious — noise, pollution, high energy use etc. — and any techniques that can throw some light on these should prove to be useful. Our patterns of energy consumption are characterised by excessive variation and instability. For example, the average American uses about 30 times as much energy as someone from the developing world. In the last 30 years we have consumed as much fossil energy as had previously been consumed in the history of mankind.

On the other hand, with many passive buildings using almost no energy at all, and photovoltaics and windpower continuing to make great progress, it looks as if a solution for the built environment could be just around the corner. It is now man's obsession with travel that we have to come to terms with. I suspect that for every business journey the Internet has saved, it has unfortunately inspired another punter to jet away on a cheap holiday flight to the other side of the world. One return flight to Australia consumes as much energy as a modern three-bedroom house in a year!

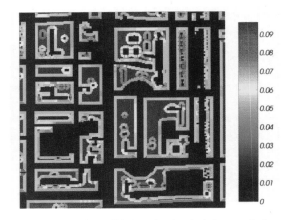

Figure 8 Output from LT Urban showing the influence of urban form on total energy use (assuming all office use).

REFERENCES

1. Martin and March. *Land Use and Built Forms*. Cambridge Research No.3 1966

2. Steemers, Baker, Crowther, Dubiel, Nikolopoulou and Ratti. *City Texture and Microclimate*. Urban Design Studies B. Univ. Greenwhich 1997.

3. Building Research Energy Conservation Support Unit (BRECSU) Energy Consumption Guide No. 10

4. Rasband and Bright. *NIH Image-A public domain image processing program*. Microbeam Analysis Society Journal No.4 1995

5. Ratti, Robinson, Baker and Steemers. *LT Urban: The energy modelling of urbanform*. Proc PLEA 2000 Cambridge, James and James 2000.

Energy and environmental quality in the urban built environment

Matheos Santamouris
*Building Environment Study Group,
University of Athens*

Cities are increasingly expanding their boundaries and populations. From the climatological point of view, human history is the history of urbanisation. Industrialisation and urbanisation have affected dramatically the number of urban buildings being constructed, and this has had a major effect on the energy consumption of cities. Today, at least 170 cities support more than 1 million inhabitants each, and estimates show that 80% of the total world population will live in cities by 2100.

Improved living standards have increased the space requirements per person. In America, between 1950 and 1990, the floor space requirements per person have doubled. There are also very important differences in floor space per person in Europe arising from social and economic differences. Moscow provides an average of 11.6 square metres net living space per person, while Paris provides 28.2, Oslo 47.2 and Zurich 50.6. The expansion of our cities thus demands more and

Energy and environmental quality in the urban built environment

Matheos Santamouris

more land to support them. In Europe, 2% of agricultural land is lost to urbanisation every ten years. An average European city of one million inhabitants consumes 11,500 tonnes of fossil fuels, 320,000 tonnes of water and 2,000 tonnes of food every day. It also produces 300,000 tonnes of waste water, 25,000 tonnes of carbon dioxide and 1,600 tonnes of waste.[1]

The direct and indirect needs for land are well represented by the notion of the 'ecological footprint', defined as the land required to feed a city, supply it with timber products and reabsorb its carbon dioxide emissions with areas covered with growing vegetation. This concept helps to set limits to the activities that an area can absorb in a sustainable way. London's ecological footprint is close to 20 million ha, which is 125 times higher than its actual surface area. In addition, increased urbanisation adds an important additional cost as new infrastructure has to be developed, and the existing infrastructure in the old parts of the city is used less. Studies in UK show that almost 60,000 hectares of land are vacant in urban areas and 15– 20% of offices are empty.[2]

The urban climate

Air temperatures in densely built urban areas are higher than the temperatures of the surrounding countryside, a phenomenon known as a 'heat island'. Factors influencing the 'heat island effect' are, among others, climate, topography, physical layout and short-term weather conditions. Data[3] show that heat island intensity can be as high as an extra 15°C.

In addition to temperature increase, the urban environment affects many other climatological parameters: global solar radiation is seriously reduced because of increased scattering and absorption; sunshine duration in industrial cities is reduced by 10–20% in comparison with the surrounding countryside, and similar losses are observed in received energy. The urban environment also affects precipitation and cloud cover. In Budapest, cloud cover has been increased by 3% during the winter period.

Energy impact

Urbanisation leads to a very high increase in energy use. A 1% increase in the per capita GNP leads to an equal (1.03%) increase in energy consumption. However, an increase of the urban population by 1% increases the energy consumption by 2.2%. Comparison of the energy consumption per capita for the inner and outer parts of cities shows that the consumption in the inner part is considerably higher. Inner London has a 30% higher energy consumption than the outer part. Buildings are the largest energy consumers in cities. In Europe, the energy consumption by the residential sector varies between 28% and 48% of the energy consumption of these cities. At the same time, buildings in the commercial sector absorb between 20% and 30% of the final energy consumption of the cities.

Increased urban temperatures have a direct effect on the energy consumption of buildings during the summer and winter periods. During the summer, higher urban temperatures increase the electricity demand for cooling and consequently the production of carbon dioxide and other pollutants. On the other hand, higher temperatures may reduce the heating load of buildings during the winter period. Taking into account that urban temperatures during summer afternoons in the US have increased by 1–2°C during the last 40 years, it can be assumed that 3–8% of the current urban electricity demand is used to compensate for the heat island effect alone.[4]

Increase in energy consumption in urban areas puts a strain on utilities. The construction of new generating plants is an unsustainable solution, being expensive and slow. The adoption of measures to decrease energy demand in urban areas, like the use of more appropriate materials, increased planting, use of heat sinks etc., in association with a more efficient use of energy, seems to be a much more reasonable option. A megawatt of capacity is actually eight times more expensive to produce than to save, because energy-saving measures have low capital investment and no running costs, while the construction of new power plants involves high capital investment and high running costs.

Air quality

Increased urban temperatures affect the concentration and distribution of urban pollution because heat accelerates the chemical reactions that leads to high ozone concentrations. In Europe it is estimated that in 70–80% of cities with more than 500,000 inhabitants the levels of air pollution exceed the WHO standards at least once a year. Comparisons of daily peak temperatures in Los Angeles and 13 cities in Texas with ozone concentrations show that as temperatures rise, ozone concentrations reach dangerous levels. Polluted days may increase by 10% for each 2.7 °C. increase. It is calculated that the cost of sulphur dioxide damage to buildings and construction materials might be in the order of 10 billion ecu per year for the whole of Europe.

Damage from increased pollutants is obvious. Analysis of the relationship between hospital admissions and sulphur dioxide levels in Athens found that a threefold increase in air pollutants doubles hospital admissions for respiratory and cardiovascular disorders. Health problems associated with the urban environment are mainly associated with the increased use of cars. This has been acknowledged recently by the British Medical Association. Pollution from gasoline and petrol has been proved to be partly responsible for heart diseases. It has been shown that, in London, 1 in 50 heart attacks treated in hospitals was strongly linked to carbon monoxide, which is mainly derived from motor vehicle exhausts.

The role and the impact of outdoor conditions on indoor climate, as well as the relation between outdoor and indoor pollution, are obvious. Intensive urbanisation and deterioration of the outdoor air creates a new situation with serious consequences for indoor environmental quality. In fact, outdoor pollution is one of the sources of 'sick building syndrome'. Increased outdoor concentrations seriously affect the indoor concentration of pollutants.

The problem of noise

Noise in cities is a serious problem. Unacceptable levels of more than 65 dB(A) affect 10–20% of inhabitants in most European cities. The OECD reports that 130 million people are exposed to unacceptable noise levels. In towns with between 5,000 and 20,000 inhabitants there are increases in noise disturbance of from 17% to 19%, while in cities between 20,000 to 100,000 it is 19% to 25%, and finally in cities of more than 100,000 it rises to between 22% and 33%.

Conclusion

Continuously increasing urbanisation, and the recent upsurge of concern for the environment and the development of new energy technologies, define the major priorities for urban buildings. This requires ideas like those developed by the New Urbanism movement, based on mixed land use, greater dependence on public transport, cycling and walking, and decentralisation of employment location. In parallel, instead of treating the symptoms of urban environmental problems, we should treat the cause: the quality of the urban

Theme 3

Testing and Modelling: the metabolism of cities

environment. Sustainable cities are cities that provide a liveable and healthy environment for their inhabitants, and meet their needs without impairing the capacity of the local, regional and global environmental systems to satisfy the needs of future generations.

Making cities sustainable entails minimising the consumption of space and natural resources, rationalising and efficiently managing urban flows, protecting the health of the urban population, ensuring equal access to resources and services, and maintaining cultural and social diversity. Environmental quality of indoor spaces is a compromise between building physics applied during the building's design, energy consumption and outdoor conditions. As buildings have a long life of several decades or sometimes centuries, all decisions made at the design stage have long-term effects on the energy balance and the environment. Thus the adaptation of existing and new urban buildings to the specific environmental conditions of cities in order to efficiently incorporate solar and energy-saving measures seems to be of very high priority, especially when more than 70% of building-related investments in Western Europe are channelled into urban renewal and building rehabilitation.

None of the above should be seen as isolated areas of concern, however. The interrelated nature of the parameters defining the efficiency of urban buildings requires that theoretical, experimental and practical actions undertaken at the various levels should be part of an integrated approach.

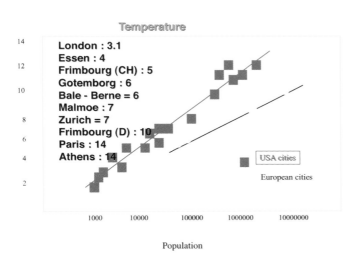

Figure 1 Temperature increase in various European and American cities as a function of the urban population.

REFERENCES

1 Stanners, D and Bourceau, P (eds) *Europe's Environment: The Dobris Assessment* European Environmental Agency, Denmark (1995)

2 Smith, M, Whitelegg, J and Williams, N *Greening the Built Environment* Earthscan Publications, London (1998)

3 Santamouris, M (ed) *Energy in the Urban Built Environment* James & James, London (2000)

4 Environmental Protection Agency *Cooling our Communities: A Guidebook on Tree Planting and Light Colored Surfaces* EPA, Washington, DC (1992)

Theme 3

Testing and Modelling: the metabolism of cities

Discussion

■ **KESTER RATTENBURY** Nick Baker, in the model you were describing, it looked as though you were going to be able to come up with an ideal form or best possible form for buildings. Is that one of the aims?

■ **NICK BAKER** It is, yes. I think the energy use in a building that you predict with a simple model is often very different from what is actually experienced. But there must be a reason for what's experienced. It is not as if it's some sort of magic. It's still obeying the laws of physics. So what we're using the model for is to try and find the degree to which urban morphology is actually determined. We get situations where, if buildings get too hot, they are air-conditioned. They then generate plumes of warm air, which the rest of the building then ingests for its packaged air-conditioning, and suddenly the energy consumption rockets. So we're dealing with very unstable situations.

■ **KATE MACINTOSH** A question for Professor Santamouris. You were very keen on decentralisation. I would like to ask whether, in your assessment of the energy consumption of people living in the country as compared with living in the city, you are taking into account transportation costs?

■ **MATHEOS SANTAMOURIS** This is very important. When the energy consumption of urban and rural buildings was compared, we

Discussion

compared just the energy consumption for heating, cooling, and sometimes for lighting. There were several experiments trying to compare the global energy consumption, including transportation, of an old non-insulated building downtown in an American city, facing the problem of increased temperatures, etc. with a new well-insulated building outside the city. The energy for transport was much higher, and they found that the energy, the global energy consumption, was almost the same.

■ MATHEOS SANTAMOURIS No, I don't think so, because when you compare the energy consumption downtown in a city and around the city, you have almost the same group of people with the same habits and the same way of life. The style of life is exactly the same in downtown Athens or in the surrounding area. So the energy consumption, the differences in energy consumption, has more to do with climate changes.

■ DAVID CLEWS (University of North London) I wonder whether you could tell me: is the increase in energy consumption in inner cities verses the outer parts of the cities, suburban parts, rural areas, entirely to do with physical conditions, or are inner city dwellers just more profligate, rather like holiday-makers with their use of energy? Is there is something about living in cities that makes us leave the lights on or leave the heating up higher?

■ NICK BAKER Could I just add that that's in an overheated city. Many cities are not overheated, and so a bit of heat-island effect in Stockholm is probably rather welcome in the winter. I think it's somewhat climatically dependent.

Theme 3

Testing and Modelling: the metabolism of cities

■ **BILL BORDASS** (William Bordass Associates) I just wondered if you could draw the line between profligacy and intensification. What's often happening in buildings these days is that people are actually using them harder. I would therefore be very careful in differentiating between these [two terms]. I looked last year at the differences between British and Swedish offices. Part of the reason why Swedish buildings use less energy is that they have about 50% more area per person. You have to be very careful about that.

■ **NICK BAKER** Yes. Maybe that is something that planners should take into consideration: you are not allowed to have a certain industry unless you find a use for your waste.

■ **MATHEOS SANTAMOURIS** I think that net watts of energy are extremely important for cities, but at the same time we have to think that by installing networks we increase the so-called ecological footprint of the city. Every year we are losing 2% of agricultural land to urban areas, and in the last century the urban land per capita has increased ten times. If you are going to use

■ **BRIAN MARK** (Fulcrum Consulting) There has not been much discussion of waste treatment in urban settings — I count rejected heat as waste — and surely the district systems that can selectively move a waste product from one user to be an input for another user must be a very important method of reducing environmental impact?

■ **BRIAN MARK** Is there a definition of wasted heat? The problem is that rejected waste heat causes a problem, whilst people are desperately trying to produce heat.

Discussion

centralised and not decentralised energy techniques, you are going to increase the ecological footprint. The ecological footprint of India is 0.16 ha per person, and that of the States is 2.06 ha to satisfy demands. Decentralisation is better from the environmental management point of view because you have less heat generated, and there are a lot of important benefits.

■ MICHAEL DICKSON (Buro Happold Engineers) I don't think you can talk about the metabolism of an urban environment without considering in full the urban footprint of that city, because the extent to which the energy you require to bring your food etc. is very much part of the impact of that urban environment on the global situation. One of the things that we need to do is think about whether there are different sizes of urban community that have a different ratio of impact. I wanted to ask Professor Baker, in your considerations of urban morphology, have you managed to study the effect of transpiration of landscape in moderating urban climate, or is that too complex?

■ NICK BAKER No, it isn't too complex, and it's very important because, of course, the heat-island effect is relative. Whenever solar radiation is absorbed it causes heating; that's what makes our weather. Surfaces like tarmac and concrete and roofing materials absorb and retain more heat than vegetation does because of transpiration. So really the heat-island effect is relative to the normal landscape, which is normally transpiring and normally vegetated. Even if it isn't, it is still losing water by evaporation, so the ratio of green space to non-green space is a key parameter.

Theme 3

Testing and Modelling: the metabolism of cities

■ MICHAEL DICKSON: So if we dug up all our boulevards and made them into landscape, would that change the temperature?

■ NICK BAKER I am sure it would. What matters is the temperature on the pavement. Local conditions may be almost more important in some respects for how people enjoy and use their city. So that's where the morphology – the microscale morphology – comes into it.

Volker Gienke, Andrew Ford, Mark Hewitt, Katrin Bohn and André Viljoen

Synthesis and Shape
designs on the city

Theme 4

Hidden technologies

Volker Giencke
*Institute für Hochbau und
Entwerfen, Innsbruck*

Weather has long served as a guide and witness of change in our everyday lives. Atmosphere may be at the core of architecture, but it is a core that cannot be addressed or controlled. The magical figure of the architect only survives in the apparent play between atmosphere and building, ephemeral climate and material object. (Mark Wigley)

Figure 1   Greenhouses in the botanical garden at Graz: the fog system.

The connection between science and architecture, in which I am very interested, is still in our minds; it has not become reality. And of course my point of view is the view of an architect, not the view of a scientist. My topic is hidden technology, which, in a very subtle but direct way, creates new architecture and new urban forms. But I also refer to the influence of weather and special microclimates.

Hidden technologies

Volker Giencke

The glasshouses in the botanical garden of Graz are a design from the early 1980s. It is an aluminium construction wrapped in a double layer of Plexiglas. What is really interesting about the glasshouses, besides the transparency, is that they simulate a different atmosphere. There is nothing but a physical barrier between outside and inside. All the installations are hidden in the construction. For instance the heating water is run through the two main tubes of the construction. Cooling is achieved by fog, a compressed air and water system plugged into the construction. When the cooling system is functioning, the houses are instantly cooled by 5°C.

In the spring of 1999 my students started to build a *Gitterschale*, a wireframe-like construction in the courtyard of our University. In the beginning, the construction looked more like a musical instrument, timber slats organized in a grid, connected and stretched into shape. Erected as a perfect shape in the morning, when the air was cool and fresh, the construction cracked and crashed to the ground in the midday sun. What had happened? The wet and pliable timber-slats dried out in the sun and broke where knots and knot holes weakened their cross-section. Invisible forces were set free. The construction became an experiment.

Seckau is a monastery in the upper part of Styria. The existing building extends into a new building, which turns inside out. One has the experience of a one-room home that changes its function from an introverted to an extroverted space. Light breaks through the cracks in the single wall – built of wood and insulated with

Figure 2 *Gitterschale*: a wireframe-like construction.

Figure 3 Seckau Monastery: choral space.

yellow glass wool – and illuminates the interior in a very subtle way.

The forces that shape

Below the surfaces of material, of architecture, and of soil, secrets are hidden. Sometimes, when I start the construction of a new building somewhere on flat land, I ask myself: who claimed ownership of a land we built on before we arrived? There is nothing special about the cornfield at the edge of the city. It bears the crops and then lies stagnant, following the seasons like all other fields surrounding it. The gravel soil makes a good building site.

The wasteful treatment of building space and the certainty that the best building site has been used for crops for thousands of years, make a large contribution to this mad reality. Nevertheless, the site is lived on. It has become a home for immaterial qualities, through its quantitatively and qualitatively meagre use. At the moment building activity starts in this place, all the creatures will flee to the surrounding rooftops and trees, and we observe the development with disgust.

It is an architect's job to create elements that stand alone like sculptures and together naturally create a space that combines the functionality of the building with the inherent qualities of the site. Specifically, architecture is a physical meeting place, hide-out or home of the spiritual, the absence of which creates our desperate urban work, or prompts us to go to the movies. If architecture happens, the original spirit

Hidden technologies

Volker Giencke

of the site is lost. It is replaced by a new one.

The history of mankind is like the history of the city: always the history of location, landscape and geography. Let us assume that the same technology, the same technological standard, and the same amount of invested capital allow the same architecture all over the world. If this were the case, the criteria that would make a location unmistakable, such as beauty of the landscape, poetics of location, and distinctive architecture, would become very significant. Under these conditions architecture is artificial environment, a work of art that takes on a special role.

Architecture must be better than what nature and history have created. The freedom of architecture is the freedom of art. Art exists independently of life's problems. It doesn't provide any solution for these problems, nor is it the solution to problems that in fact never existed. Art can only be radical as an illusion. The dimensions of sensuality, fantasy and intellectuality are complementary to the daily grind of society as we know it today. The revolutionary potential of art demonstrates that it can only be art that will be able to satisfy the non-materialistic needs of mankind. I consciously compare this to religion, because man's destiny is to fight against his destiny.

Innsbruck and the 'Viech'

A few words about Innsbruck: approximately 200 million years ago the history of the Alps starts with a crack in the ancient continent. Only over the last 50,000 years have humans played a role in this world. Innsbruck is a thousand-year-old protected and threatened by the Alps. Organisms adapt constantly through evolution and, as a result, form an ecological system. The 'Viech' is a project that has hidden itself in the hills of Rum and Thaur, which are small villages near Innsbruck. As an instrument with the behaviour of an animal the Veich is waiting to be fed by the farmers. Like an organism it transforms its food into biogas and dies. The biogas is stored in a bubble, which also acts as a storage place for its wings. When the sun is shining, the bubble grows; it opens when the gas pressure becomes too high and stretches its wings towards the light. When the sun disappears, the wings recede into the public end. The power plant is a biogas plant, forming a functional unit with foldable solar collectors. A mechanical 'sun guard' moves the rail on with the solar collectors waiting in a folded state. The rain increases the biogas bubble and triggers its operation. Only the blunt form of the bubble allows the collectors to move out. If the bubble is slack, it locks the opening and encloses the collectors in the hill.

Detroit and/or sustainability

The town planning and urban design of today mean to me, in an ironic way, how to create the ugliest city in the world. Boredom, ugliness, roughness are criteria. A city has to bear what is ugly; a city is a mixture of chocolate cream and tomato ketchup.

Hidden technologies

Volker Giencke

Of course a city is not a dream and not a homely house... but it is a home, a place to live, a place to be. I quote Cedric Price: 'A city can be wonderful if you are driving through it and it can be horrible if you have to die in it ...' This sentence should qualify every effort in urban design.

The 'disappearance' of Detroit in the 1970s, one of the most modern cities in the world, what we call the 'Motor City' fascinates me, as it frightened the urban designers. It shows that there is nothing built sustainable in terms of history. And it is a big chance for a new Detroit, a new idea of urban environment, a kick in the butt for every boring urban designer. Sustainability as an urban conception works against flexibility, against change, against the liveliness of transitoriness, against the rhythm of life. Sustainability is nothing else than a fashion of our time – not at all the answer to dislocation and timelessness.

Emerging city shapes: energy at the urban scale

Mark Hewitt
d-squared design/University of North London School of Architectural & Interior Design
Andrew Ford
Fulcrum Consulting

We know that we need to make cities healthier and more energy efficient. We also know that, to achieve this ambitious aim, the skills of many different kinds of people are required, including transport specialists, policy thinkers, planners, physicists and cultural commentators. Designers are a small part of the complex weave that generates a city. Nevertheless, being designers ourselves, we identify a need to focus our attention on developing an approach to designing more sustainable cities. Over the last three years we have been working with groups of post graduate architects at the University of North London, and environmental engineers from Fulcrum Consulting, in a collaborative attempt to develop such an approach. The collaboration is important since it is our view that only through a much more sophisticated understanding of the fields traditionally occupied by the respective disciplines of the architect and engineer can progress in sustainable urban design occur. Our project has

Emerging city shapes: energy at the urban scale

Mark Hewitt and Andrew Ford

been to establish a shared understanding of both the spatial and physical dimensions of city performance in order to create a synthesis of architectural and engineering skills. In essence, our three-year project has been to design an approach that can evolve viable urban patterns from the social and ecological realities of place.

Mappings

Our work always begins with looking at a city, through a series of different analyses or 'mappings'. Some of this work is quite speculative. In one example, a group of London and Innsbruck students looked at the energy potential of an area of open ground in Innsbruck, asking some elementary questions about the relationship between the energy that might be derived from organic growth compared with building structures over a period of 25 years. This kind of study provokes some strategic questioning, particularly about how urban land use is traditionally allocated. Often the use categories and densities of an urban sector are designed with reference to traditions and norms, without any 'first principles' thinking about where these received models come from.

There is a big idea behind all of this: a lot of solar energy falls on any square metre of the earth and for any given place we know how much. In a typical European city there is enough solar energy to provide for all of the energy requirements of the buildings and their inhabitants, if you build right. Sustainable cities (in energy terms) are therefore

Figure 1 Inn Valley Halle region energy analysis: wood plantation.

Figure 2 Inn Valley Halle region energy analysis: elephant grass plantation.

Figure 3 Inn Valley Halle region energy analysis: photovoltaic array.

technically feasible even within current capabilities, but we need to set much more ambitious targets and allow ourselves room for imagination. In analysing the remaining open ground between Innsbruck and Halle in the Inn Valley we considered various ways in which energy might be harvested. If the area was planted as forest, it could produce 21 million kWh per annum. This is the amount of energy you would get if you burned the annual wood crop. We could cover the entire area with solar panels, a disturbing thought but a very current proposal for European cities. If you did that on our site you would get 181 million kWh per annum. If you planted the area in elephant grass, you could produce 33 million kWh per annum. If you covered the area with buildings, you would get a large deficit: how large would depend on the efficiency of the construction.

The concept of the city as a modifier of environmental conditions has an ancient history. Vitruvius addresses this issue first, before even tackling the questions of defence or water, and counsels against ignorance of the urban microclimate: 'For heat is a universal solvent, melting out of things that power of resistance, and sucking away and removing their natural strength with its fiery exhalations . . .'.[1] The problem for Innsbruck is not necessarily heat (although many of the city's offices are now air-conditioned), but the potential for the city to be both a positive and a negative environmental modifier remains. We have found that one of the best ways to begin to understand how cities work is to make a 'snapshot' through mapping several conditions simultaneously. The method is

Figure 4 Neutral buoyancy balloon release exercise in Innsbruck to track air currents in dense urban, suburban and open field situation.

Figure 5 Agricultural area remaining between Innsbruck and Halle.

Emerging city shapes: energy at the urban scale

Mark Hewitt and Andrew Ford

based on direct measurement and observation: how do air currents flow in different urban densities? What is the thermal performance of different parts of the city at both a building and a district scale? In a study of air movement, a group of students made a simultaneous release of balloons in three different parts of the Inn Valley – dense urban, suburban and open farmland – and then tracked their movement to get a snapshot of the complexity of air currents.

Such effects are often so complex that they are not intuitively obvious to designers. Even for expert building physicists, modelling such effects is a specialist research activity, which perhaps is available only for specialist or research projects. And yet, as the work of Professor Baker and others has shown,[2] quality of air in a city can be hugely variable even in a local area, depending on the particular flow patterns. In the Innsbruck study, the tracking of air currents helped to reveal a local anomaly: inversion currents, which cause polluted air to get trapped at around 680 metres within the bowl of the valley. This is the zone that is currently earmarked for further residential development, so that despite the fact that it is above the downtown area of the city, and has wonderful views of the alpine scenery, the zone turns out to be quite an unhealthy location for dwelling.

Another mapping technique, thermographic imaging, provides information about how heat flows operate in the city. Often this is neither obviously apparent nor intuitive. Through considering and monitoring these specific effects, designers can build up a richer sense of what particular fabric elements can do, in terms both of the 'architectural' concerns of colour, surface, texture and weathering, and of the

Figure 6 Thermographic study of the central part of the city of Seville.

Figure 7 Thermographic study of building fabric in the central part of the city of Seville.

'engineering' concerns of thermal capacity, reflectance, and surface emittance characteristics. In this way a more sophisticated discussion can evolve about possible choices within the design group. What do the materials that make up the urban fabric really do? Students were able to use this information to refine design choices, for example using frozen sheets of water in winter to reflect low winter sunlight deep into a volume. In Innsbruck we observed how the heat of the dense urban centre gradually diminishes as you move towards the more dispersed periphery, and – at a more detailed scale – how individual fabric elements that make up the buildings of the city have very different thermal performance characteristics.

At the urban scale, a knowledge of solar geometry and orientation is critical. In the Inn Valley study it became apparent that the northern mountain ranges are in fact the key shading elements in the summertime, not the southern mountains. This is counter-intuitive, and the result of a combination of mountain topography and sun angles that results in the casting of morning and afternoon shadows on south-facing slopes. Such information may be a critical factor in deciding, for example, which areas should be premiated for agricultural production and which for built fabric. There are other categories of 'mapping' that we include in our methodology,[3] but there is only space to touch here on a few.

Proposal and simulation

Having built up a sufficiently clear picture of the complex interactions, both social and physical, that

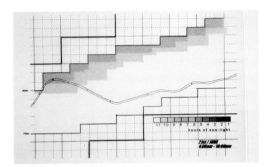

Figure 8 Graph showing summer sunlight hours distribution in Inn Valley. Midsummer. North at top of image. Eva Benito Benito, Unit 6.

 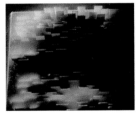

Figure 9 Photographic heliodon study of shadow distribution in Inn Valley. Midsummer. Eva Benito Benito, Unit 6.

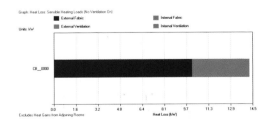

Figure 10 Plot from study of energy flows in a small building. Student of Unit 6 using Integrated Environmental Solutions "Facet" software.

Emerging city shapes: energy at the urban scale

Mark Hewitt and Andrew Ford

characterise a city, we can feel more confident about the ground upon which design proposals are made. City patterns should be evolved from an understanding of performance. On our course we do not yet make digital simulations of the performance of urban scale ideas; however, we can make simulations at the building scale using Integrated Environmental Solutions software in the envirolab.[4] Once students have gained proficiency with these tools, they can apply this to their own design projects. Simulation of building fabric performance is a useful way to test out variations of ideas. For example, in a project by Paul Saltmarsh, the programme of the building, a community hall, is tested against different material manifestations to see how the placing of openings and choice of materials interact with the use pattern. The amount of solar energy potential for capture and storage within the building fabric was simulated. It became clear that the envelope of the building received more than sufficient energy in the summer months to cover the winter heat requirement. This inevitably provoked a study of whether it was possible to develop a strategy for transferring the energy surplus across the season.

In the resolution of this particular project, a basement was used to store the summer heat in large tanks of water for subsequent air heating by conduction. The labyrinthine pattern of the tanks is designed to maximise surface contact with the passing incoming air, and also to make a moody entrance for the nightclub above. Simulations of the amount and path of the airflow help to refine the structural design of the project, provoking a

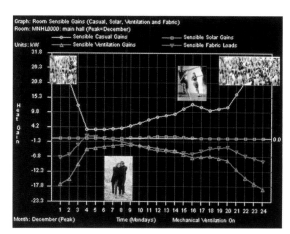

Figure 11 Heat gain study to assess heat load from inhabitants of a nightclub, using Integrated Environmental Solutions Software, graphically enhanced. Paul Saltmarsh, Unit 6.

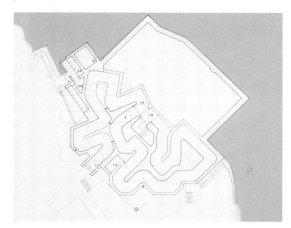

Figure 12 Basement labyrinth heat store for summer energy reuse in winter. Also providing moody entrance sequence to a club. Paul Saltmarsh, Unit 6.

Theme 4

Synthesis and Shape: designs on the city

cambered, coffered structure to allow the natural buoyancy currents to remove exhaust air. By working between simulation of performance and invention of form, a more tuned architectural project can begin to emerge.

The use of performance simulation in the envirolab at the University of North London is an essential part of this approach, since it is one clear point at which the aspirations and backgrounds of the architect and engineer can be fused. By working on a design even in the relatively abstract and simplified geometry of a simulation programme, the architect and engineer can discuss the implications of decisions as they occur. They avoid the non-simultaneous pattern of proposition, usually by the architect, and calculation/response usually by the engineer, divided by a gap in time and understanding, which is the normal experience.

Simulations at an urban scale are a complex problem and outside the scope of our current project, although we have begun to consider ways of moving in this direction. Real-time observations of air movement gave us a snapshot of some of the air movement characteristics for the city. In Innsbruck there is a particular problem with inversion currents, as we have discussed. This would be a good subject for simulation; we have identified the problem through observation, but are unable to test our propositions for change. At the moment, the design proposal in response to pollution is to develop a planting strategy that will provide increased air purification, and the channelling of stronger winds through belts of trees to clear pollution from the

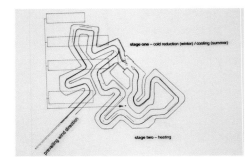

Figure 13 Air flow pattern in labyrinth heat store (see image 12).

Figure 14 Photographic study of community hall ceiling, with structure designed to encourage ventilation air flow to outlets. Outlets also provides light to hall, which is tested for change in illuminance due to future forest canopy growth.

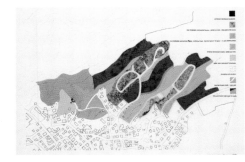

Figure 15 Site plan showing forest plantation in strips to encourage cleaning and funnelling of air currents. Paul Saltmarsh, Unit 6.

Emerging city shapes: energy at the urban scale

Mark Hewitt and Andrew Ford

inhabited areas. It would be valuable to test the efficacy of these ideas.

Mixed media for envisioning

Our project over the last three years has also focused on the importance of the representation of urban ecology. This is a matter of envisioning, and in our minds it is impossible to think about such complex conditions unless there are ways of representing them. The model that was constructed from the mapping work done in Innsbruck is an attempt to represent some of the main issues, and we are calling this approach 'topo-energetic' modelling. The issues include pollution concentration, how a city values and defines its agricultural hinterland, the potential for using ground heat energy at an urban scale, the establishing of a density ratio between organic growth and built fabric, and non-polluting infrastructure networks. Because all of these by definition deal with forces and quantities that are moving in time and space, the topo-energetic model uses light to represent critical shifting forces. For example, the shifting pattern of meltwater drainage streams, which swell seasonally to feed the extraordinary torrential force of the topaz-green River Inn, is represented by thin traces of photoluminescent paint. When illuminated with ultraviolet, the network of streams become apparent. The activated lines glow brightly for a period of time, representing their presence in the annual cycle of the Inn Valley, and then fade out, as the streams become dry for the remaining months of the year. This model

Figure 16 Innsbruck street.

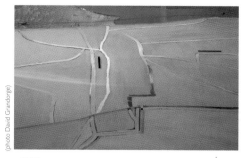

Figure 17 Topoenergetic model: detail showing spring meltwater streams. Unit 6.

of representation is capable of describing movement and transformation in time, essential for revealing the dynamics of forces such as heat and light, and changes in topography and urban form.

Conclusion

We believe that by finding ways to represent some of the complex dynamic forces that provoke the evolution of cities, we can start to evolve viable and intelligent urban patterns from the social and ecological realities of place, and also begin to engage the imaginations of designers in these issues.

REFERENCES

1. Vitruvius *The Ten Books on Architecture* (translated by Morgan, H) Dover, New York (1960)

2. Baker, N et al. 'The urban porosity model: simplified parameters as indicators of environmental performance' *Proceedings of Fourth European Conference on Architecture*, H.S. Stephens and Associates, Berlin, 1996 pp 300–305

3. Hewitt, M, Ford, A and Ritter, D 'Using design imagination to develop sustainable urban strategies' *Proceedings of PLEA 2000 Conference,* James and James (Science) Publishers, Cambridge, July 2000 pp 702–707

4. 'Envirolab' is a building performance simulation laboratory set up in 1998 at the School of Architecture and Interior Design at the University of North London, in collaboration with IES Software Ltd and sponsoring partners from industry

Figure 18 Topoenergetic model: detail showing computer controlled ultraviolet light sources for illustrationg dynamic shifts. Unit 6.

Figure 19 Topoenergetic model: changing landscape between Innsbruck and Halle. Unit 6.

Figure 20 Topoenergetic model: before sequence of proposed changes. Unit 6.

Discussion

■ **NICK BAKER** I think the argument about the amount of solar energy that falls on to a city, and the amount that you need, is very persuasive. But unfortunately all energy is not equal, and even growing plants only work at about 1 or 2 % efficiency at the best. So it's not easy to use energy in the way we want to. There has been a long history of thermal storage projects, but they're a bit iffy and, unfortunately, extremely uneconomic at the moment. So without trying to put it down, the project has got to be placed in the context of many efforts for long-term thermal storage, which doesn't, after all, store electrical power for lighting.

■ **ANDREW FORD** I accept that, but if we can get rid of the thermal problems, we have dealt with a lot. The thermal issue is critical to this whole debate. We are trying to achieve comfort with what's available. In some places you have 'cold' available, in some places you've got 'hot' available, so ... Storage is the key. In this part of the world we have got more cold available. So in many ways what you are trying to do is to design things so they just need cooling. It would be easier.

■ **ROGER BURTON** Again, some of these strategies have been around for a long time. I seem to remember starting my career in the 1970s, when everybody was thinking about low-energy housing, very highly insulated housing, and how you could store heat. A lot of interesting and intriguing pieces of architecture were made. The problem has always

Theme 4

Synthesis and Shape: designs on the city

been: how do you make people take up these strategies? They have been around for a long time, and people still air-condition their buildings. If it is really going to be a workable urban strategy, there is something missing. I am sure it is actually understanding that there are environmental consequences to the use of energy, and pricing those appropriately. Until we do, people will carry on doing what they're doing.

■ MARK HEWITT A village hall.

■ NICK BAKER I think the problem is that it's a piece of architecture, and the notion of storage, thermal storage, works better the greater the scale. I honestly think that without that economy of scale of storage, it wouldn't work. I know there have been enough monitored indexes on storage projects to know that on a single housebuilding basis, there is almost no hope.

■ VOLKER GIENCKE Yes, I think in general I have to say that it is very strange for me that your general urban concept consists only of statistics and diagrams. I think there is a prerequisite for an urban environment that also is an organism. It is also a prerequisite for architecture. The last example you showed is a small hall.

■ VOLKER GIENCKE I think it was very convincing because the basement really follows directly your consideration about heating and air ventilation, and so on. But there is something that has to do with how you play with these considerations, and that's what I am talking about when I talk about a new city or a new architecture.

■ ANDREW FORD I agree.

Discussion

■ NICK BAKER But on an urban level it might work, so it's an idea we should consider.

■ NICK BAKER It's important to remember the behavioural factors in energy needs, because a person in the West consumes energy at the rate of about 7,000 watts per day, whereas a person in India consumes it at about the rate of 90 watts per day. Some days we probably use no energy at all and on other days we use huge amounts because it's part of our lifestyle.

■ HELEN MALLINSON (University of North London) There is an equivalent in my mind between the separation you were talking about in terms of economic and environmental sustainability, and the separation of our professions. This notion of the diffusion of separate cultures applies equally to the way we work as professionals. In actual fact, diffusion will not occur in practical terms until it occurs within education and the professions. I think that someone said earlier that we know too much. It's one of our problems. Communication between our disciplines is the key issue of the moment. To me that's the primary issue.

■ DAVID CLEWS I think there's a very powerful tool being developed here. I think it is admirable the way the physics of all of these problems has been addressed. Architects' lines of communication aren't the same as they are for physicists or scientists. So to try and mediate and find a language between building physics and design seems to be bearing some fruit.

■ DAVID RITTER In terms of the debate about urbanism, which is what this is about really, I think there is a danger that you can overlook the separation between economic sustainability and environmental or ecological sustainability. I think proposals like these actually reinforce the link between the two.

Response to Theme 4

Katrin Bohn and Andre Viljoen
Bohn and Viljoen Architects
Andre Viljoen is Deputy Director of the Low Energy Research Unit at the University of North London

Figure 1 Sustainable cities, new landscape elements.

Figure 2 Sustainable cities, new landscape elements.

Architects

This theme's two papers confront us with sometimes opposing views on sustainability and the role of today's architect within it. Mark Hewitt and Andy Ford discuss new possibilities arising out of the collaboration between architects and other professions to design sustainable cities. They are trying to find ways of envisioning temporal human and environmental interactions, 'which can evolve viable urban patterns from the social and ecological realities of place'. Volker Giencke suggests that sustainability may limit the artistic scope of architects, that it may destroy creativity and poetry, which is often driven by uncertainty and unbounded possibilities, that 'the magical figure of the architect' may get lost.

So, limitless possibilities versus narrow horizons – is the role of the architect diminished if ecological and social concerns seriously influence design decisions? Or do architects and architecture gain through the integration of new

Response to theme 4

Katrin Bohn and Andre Viljoen

Figure 3 Sustainable cities, testing: A 370m length of the Holloway Road, in London, analysed to find out the area of mini market garden required to supply its annual fruit and vegetable requirements.

and broader aspects of life? Is collaborative working within the built environment limiting, or is it a creative response to contemporary conditions? Does sustainability increase or reduce the originality of architecture? Or, to put the questions differently: should sustainability provide the context for architectural endeavour?

Architecture

Environmental change of significant dimensions appears to be with us. We know all about it: carbon dioxide emissions, resource depletion, acid rain etc. will seriously disrupt the status quo on Earth. This has already resulted in initiatives to work within the limits of natural renewable systems and to confront the exhaustion of world resources. Sustainability has become the word to describe these initiatives. Sustainability is large. One knows, at least roughly or intuitively, what it means. It is one of the largest views of life on Earth, in some ways comparable to Greek philosophy, ancient Buddhism or Kant's *Humanismus*. It is about the whole world: its history, its present and its future. It includes everything: all populations, their actions and achievements, all animals, all vegetation, the four elements, Austria and the Alps.

Looking at architecture or urbanism in our environment, sustainability at any scale, whether the domestic, urban or global, has only started to be of significance. A much wider field lies before us, inviting all kinds of different people and professions to contribute. Reality is waiting to be transformed.

Theme 4

Synthesis and Shape: designs on the city

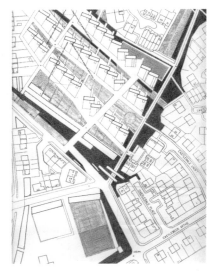

Figure 4 Elasticity, Sheffield. Continuous landscape: a proposition for connecting existing parcels of open land. These connections form ecological corridors and a landscape amenity. Proximate agriculture integrates mini and SUPER market gardens in the continuous landscape, for food, work and pleasure. These produce indigenous crops commercially for local consumption.

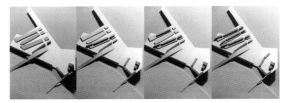

Figure 5 Urban agriculture and infrastructure, elasticity in Sheffield. A proposal for accommodating incremental development.

This can be very challenging for designers and planners. But there is a stigma attached to the sustainable city: it might be boring, inflexible, too homely, dull. Why should it? There is no reason why the sustainable city should lose the cultural vitality or identity it had before. It will change, of course, but change is in the character of cities.

One could imagine the changed city accommodating new landscapes that hold park and agricultural elements, and so a renewed connection with seasonality and materiality. Or one could envision a sophisticated state-of-the-art public transport system improving all kinds of movement through the changed city, and reducing traffic jams, air and noise pollution. Or one could see the city changing its density: underused inner-city spaces, brownfield sites, left-over plots will be occupied by new-low energy architecture individually designed to suit the requirements of the place. Or... Contemporary architects should adapt to the problems of their time. What was possible 100 or 80 years ago is perhaps not possible any longer or is not meaningful any more. Nowadays it is no use being artistic or working only on personally satisfying projects, while economists, private investors or bureaucrats plan and build the cities of tomorrow.

Art

There is a second stigma attached to sustainability: it might be something for the scientific architect but not for the artistic architect. If architecture is art (as the rumour still goes), there is no reason

Response to theme 4

Katrin Bohn and Andre Viljoen

Figure 6 Elasticity, Sheffield, landscaped fingers run between dwellings.

Figure 7 Individual sites are located on terraces located in relation to topography and solar access.

why sustainable architecture should not be the same. If architecture is not art (which we are beginning to suspect), sustainable architecture would still be more likely to be art than most contemporary architecture. It works on similar principles: it poses broad questions about our time; it seeks a vision of the future, and it can be radical, even dramatic.

Just imagine if one could call sustainable architecture art! That would attract the greatest numbers of architects. Why? Did they miss the job they really wanted to do?

There is, in our view, some confusion about the role of the architect as artist. Art does not exist in a vacuum; it responds to a contemporary condition. Good art responds in a meaningful way. Paintings do so, and films, books, plays, chansons and rap music. Architecture does (or should do) the same: it responds. In this respect the two are very similar. If one believes that artists/architects are the antenna of society, it should be no surprise that new work is not valued when first proposed. Is this what is happening to sustainable architecture?

City

Maybe planners and architects should keep their eyes closed and dream of car-friendly industrial/commercial cities like, let's say, Detroit. This has undoubtedly a certain fascination. But it seems to us not only challenging but also important to try reestablishing a liking for the city as contemporary living place and recreating responsive suburbs.

Theme 4

Synthesis and Shape: designs on the city

Figure 8 Spine dwellings provide accommodation for two persons. A plot to the south can be used as a garden, or for various combinations of living/working accommodation. The plot can become the house.

Figure 9 Elasticity, Sheffield: Interior view from spine dwelling into south facing plot. Continuous landscape at the scale of a dwelling.

Figure 10 Elasticity, Sheffield: External view of houses and market gardens.

Each of two housing schemes we have designed responds to its specific city and site, both in formal and in functional terms. Each attempts to achieve sustainability within the urban environment, but at different densities and different scales. One is low-density housing (Sheffield, 72 persons/hectare), the other is a high-density development (Shoreditch, London, 450 persons/hectare); one is in a socially deprived suburban housing area (Sheffield) and one in a better-off city location (London).

In Sheffield, we explored low-energy, low-density housing within a 'continuous landscape' that accommodates parks, pleasure and productive elements. The scheme centres around the house and its relationship to, and impact on, the local environment. One new aspect of the proposal is the ability of the houses to increase in size on a given plot and to reduce if occupation changes. A second new aspect is the way productive agriculture is brought into a housing environment. Mark Hewitt and Andy Ford discuss the importance of visually representing urban ecologies as an aid to demonstrating environmental interaction, so we shall concentrate on the second aspect of our proposal as it makes interaction with the environment visible.

Taking topography and the neighbouring housing context as starting points, a landscape strategy has been developed that accommodates environmental concerns and gives meaning to the location. At the centre of the sloping site, a series of terraces are proposed that accommodate housing or market gardens. The market gardens here form a kind of private space to the houses in that they lie

Response to theme 4

Katrin Bohn and Andre Viljoen

Figure 11 Urban Nature – High density development in Shoreditch, London

adjacent to the housing plots, mirroring the ecological footprint of every household.

We call the market gardens mini- or super-market gardens depending on their location within housing clusters or in the wider area of the interconnecting continuous landscape. They would be operated commercially by professional gardeners and their organic produce sold locally. In this way, they also reduce environmental degradation resulting from remote industrialised agriculture, which is estimated to be equal to the environmental degradation resulting from energy consumed in the home and by a private car[1]. Beyond its utilitarian aspect, the continuous landscape, especially with its market garden elements, begins to make the world comprehensible. To our mind, this is important and necessary to counteract the increasingly superficial view of the world promoted by much 'theme park' culture.

In London, a competition proposal for a site in Shoreditch, abutting the City of London, tests ideas similar to those in Sheffield, but for people working in one of Europe's most important financial centres. The site is located above disused 19th-century railway arches. Two slender towers, designed to low-energy standards, provide the principal accommodation, while one-to three-storey low energy perimeter housing and workshops frame two large fields containing market gardens. Within this 'urban nature' proposal, the desires for a link with nature are related to strategies for urban sustainability. From the buildings, people overlook the seasonally changing market gardens. A green

Theme 4

Synthesis and Shape: designs on the city

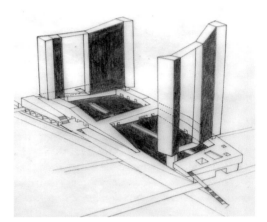

Figure 12 Urban Nature Shoreditch: Fruit and vegetable fields are placed on vertical and horizontal fields. This urban farm can produce one third of the fruit and vegetable requirements for the site's population.

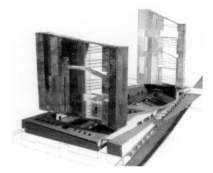

Figure 13 Urban Nature Towers, Shoreditch, Urban Agriculture in the city. This vertical housing is made readily accessible by a series of leisure walkways which zig zag up the towers. These walkways create a green complex of visible and therefore more safe public circulation.

promenade runs over the perimeter housing, connecting to the existing open land in the area around the site. As well as providing ecological corridors within the city, these green links form an alternative route through, and reading of, the city.

In part the proposal for urban nature comes from the observation of the gradual migration of people out of city centres into 'nature'. In the United Kingdom, those who can afford it, wish to live on the edge of city in the countryside. In Germany the trend is similar, with people favouring detached houses in forested areas. In neither case are people moving into 'real' countryside or 'real' forest. They'd rather position their houses in an area that provides desirable elements of nature within a suburban setting without the rest.

Common to both the Sheffield and the Shoreditch proposals is the use of landscape strategies for structuring and linking disparate parts of the environment. They become examples of how current architectural debates may be allied to environmental concerns. Discussions about infrastructure and the contemporary city are well known. New in our two proposals is the way in which locally bounded interventions at the scale of landscape may give measure and meaning to the increasingly diverse and spreading contemporary city. We can begin to envisage a naturally based urban environment within which specific areas have particular qualities and functions, and where the boundaries between those areas become places of interchange and richness through interference.

Response to theme 4

Katrin Bohn and Andre Viljoen

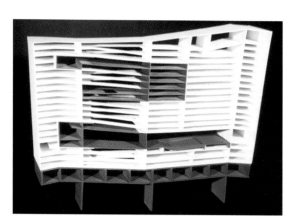

Figure 14 Vertical intensification, Urban Nature Towers, Shoreditch. Public activities normally found at street level are drawn upwards into the core of the towers. Libraries, auditoriums, nursery schools and sports facilities are stacked vertically so that high rise residents are brought into close proximity with public life.

Visions

There is at the moment a vision of the conquest of space, of Mars. Mankind works to make Mars, which is unhabitable, habitable, while at the same time making Earth, which is habitable, unhabitable. . . Why not try a sustainable vision? Perhaps there are no examples of truly sustainable building in history, or no models of sustainability left in our time. Is this of importance? Mark Hewitt and Andy Ford have referred to the complexity of simulating environmental conditions at an urban scale, but concepts developed as urban strategies can often be tested at a smaller scale.

Ecological intensification is an idea we use to test our proposals. This refers to the notion of increasing the use of renewable natural resources (sun, rain water, the soil for crops), but also to the intensification of human use, that is occupation. Increasing the density of occupation will reduce, for example, reliance on the car, the use of energy and building materials or the need for building on existing open space.

At a small scale, ideas of urban intensification are being tested in a proposal for a new live/work dwelling situated in a Victorian residential area of North London. A gap left over between two terraces provides the site: 3.15 metres wide and 20 metres long. Apparent restraints result in the compact but spatially varied 'Slot House', containing south-facing double-height living spaces, roof gardens and a ground-floor working space overlooking a walled garden. The dwelling's green roofs form a small green lung in the city, and, by

Theme 4

Synthesis and Shape: designs on the city

Figure 15 Slot House – Urban Intensification in North London

Figure 16 Slot House London, south facing facade overlooking planted roofs and designed to accept and control direct solar gain.

being raised, are exposed to the sun, stimulating growth and the potential for biodiversity. The ends of adjacent terraces provide thermal protection, while the south-facing facade assures passive solar gain. The Slot House reduces heat loss and thus the environmental impact of its neighbours, has virtually no heating requirements of its own and, by providing a new dwelling, will protect an existing piece of open land.

For most countries of the world, small-scale intervention has started. One could carry on with the larger. This would not be about providing each other with technology or invested capital, but with a vision of the future. Sustainability is the only vision of the late 20th century worth taking into the 21st. But what if even this vision proves to be wrong? We would, in this case, have to assume that the predictions on which sustainability is based, i.e. those regarding climate change, resource depletion and reduced biodiversity, are wrong. How badly off would we be having followed the sustainable philosophy anyway? We would have energy from renewable resources, increased biodiversity and less pollution of air and water. We would have stopped the spread of the suburbs into every landscape, have a better balanced transport system and more green in the city. We would have fewer health problems due to a polluted environment or toxic nutrition. And finally, if the ethics of sustainability were included as well as the physics, we would have more local jobs more equally paid and a better world distribution of resources and energy.

Response to theme 4

Katrin Bohn and Andre Viljoen

Figure 17 Model of the three metre wide Slot House, London (photo M. Dunseath).

REFERENCES

1   Vale, B and Vale, R 'Building a sustainable community' *Homes for an Autonomous Community* GIR 53, Department of the Environment, Transport and the Regions, London (1998)

Participants in discussion texts

(Bold print indicates contributors of papers)

ANTHONY AUERBACH	Vargas Organisation
NICK BAKER	Martin Centre, Cambridge University
KATRIN BOHN	Bohn and Viljoen Architects
BILL BORDASS	William Bordass Associates
ROGER BURTON	Taylor Young Architects, Manchester
JAMES CAIRD	Architect and Head of Planning for South Shropshire District Council
IAN CHRISTIE	Deputy Director, DEMOS
DAVID CLEWS	Head of Diploma Architecture, University of North London
MICHAEL DICKSON	Buro Hapold Engineers
ANDREW FORD	Fulcrum Consulting
MATTHEW GANDY	Department of Geography, University College London
VOLKER GIENCKE	Architect, professor at Institut für Hochbau und Entwerfen, Innsbruck, Austria
SUSANNAH HAGAN	Head of MA Architecture: Sustainability, University of East London
MARK HEWITT	Partner in d-squared design, Diploma design tutor, University of North London

Participants in discussion texts

STEVEN JOHNSON	Architecture Ensemble
KATE MACINTOSH	Architects and Engineers for Social Responsibility
HELEN MALLINSON	Head of School of Architecture, University of North London
BRIAN MARK	Fulcrum Consulting
MARTIN QUICK	Architects and Engineers for Social Responsibility
SUSHEEL RAO	Building Research Establishment, Garston
KESTER RATTENBURY	Journalist and writer, Symposium moderator
ERNST REINHARDT	Head of Transport Section, Energy 2000, Zurich
DAVID RITTER	Fulcrum Consulting
MATHEOS SANTAMOURIS	Building Environment Studies Group, University of Athens
ANDRE VILJOEN	Deputy Director of the Low Energy Architecture Research Unit of the University of North London School of Architecture and Interior Design
PETER WEIBEL	Chairman of ZKM Centre for Art and Media, Karlsruhe, Austria
MANFRED WOLFF-PLOTTEGG	Architect

Index

air pollution 60
    quality 83
algorithm 15, 16
artificial intelligence 17
Athens 89
Australia 78
autocatalytic models 19

bagel city 4
Baker, Nick 103, 108
Barcelona 7, 27, 28
Basle 46
Beijing 16
Berlin 62, 75, 77
Bern 46
Best Value 43
biodiversity 31, 120
Bombay 26
bottle banks 62
Bourceau, P 87
Broadacre City 8
brownfield 24
browsing 19
Bruno, Guiliana 57, 66
Brussels 19
Buddhism 113

Cairo 16, 41
Calthorpe, Peter 11, 12, 14, 22
Cambridge 73

carbon dioxide 83, 113
    monoxide 84
car pooling 53
    sharing 45, 46, 47, 48, 53, 54
Castells, Manuel 10, 14
Catholic viii
cellular automata 18
chaos theory 17
CIAM viii, 16
cities of consumption 5
City of Tomorrow 8
climate change 21, 89
cloud cover 82
Colosseum 18
computational fluid dynamics (CFD) 75
conservative 22
continuous landscape 116, 117
Copenhagen 26, 27
Le Corbusier 3, 8
CPU 17
cyber-future 9

daylight 77
DEMOS 58, 63
density 28, 52, 107, 114, 115, 119
Detroit 98, 99, 115
dioxins 60
Dunster, Bill 13, 14

ecological footprint 81, 90, 91
    intensification 119
    sustainability 111
    urbanism 58
economic sustainability 111

Index

ecopolis 38, 44
Edinburgh 55
Einstein 8
energy
    consumption 82, 83
    ground heat 107
    renewable 11
envirolab 105, 106, 108
Etoile 16, 27
Europcar 47
evolutionary models 18, 19

fantasy 97
Fishman, Robert 9, 14
fog 95
fossil energy 78
fractal geometry 16, 17
Frazer, John 18
fuzzy logic 17

gardens, market 116, 117
Garnier, Tony 3
Gence, Charles 22
Germany 118
Glasgow 55
glass 62
glasshouse 95
globalisation 10, 61
global warming 21, 31, 77
Graz 95
green consumerism 65
Greenwich 53
Gruber, Andi 18
Grune Punkt 64

Hall, Peter 10, 14
Halle 102
Harvey, David 67
Haussmann, Baron 27, 73
heart disease 84
heat island 75, 81, 82, 83, 89, 91
    sinks 83
Heavenly City viii
Hertz 47
Hopetown 13
Horton, Stephen 57, 66

IES Software Ltd 108
incineration 59, 60, 62
Inderbitzin, J 50
industrial city viii, ix
Industrial Revolution 3, 33
information city viii
    technology 33
Innsbruck 101, 102, 103, 107
Internet 23, 78
inversion currents 103, 106

Kant 113

landfill 59, 60, 61, 62
landscape 13, 114, 116, 120
land use 33
Langton, Chris 18
Las Vegas 5
Levett, Roger 41
Local Agenda 21, 43
local government 59, 60, 64
location theory 32

London 28, 40, 75, 77, 81, 101, 116, 117
Los Angeles 84

mapping 101, 102, 104, 107
March 73
market gardens 116, 117
Mars 119
Martin Centre 73
meshworks 9
Mexico City 16, 27
microclimate 70, 72, 94, 102
modelling, topo-energetic 107
morphing 20
Moscow 80
Motor City 99
Muheim, Peter 50
Murray, Robin 39, 58, 66

natural ventilation 75
Net society 5
networking 19
New Mexico 26
New York 26
Norway 75
Novak, Marcos 18

Oslo 80

Paris 16, 27, 73, 80
photovoltaics 78
planning 33, 38, 42
 town 35, 46
  urban transport 46
Plexiglas 95

pollution 10, 73, 77, 83, 84, 106, 114, 120
post-industrial city 5
Price, Cedric 99
public transport 11, 26, 46, 47, 49, 114

Quaker viii

recycling 39, 41, 44, 57, 58, 59, 60, 61, 62, 63, 64, 65
reflectance 104
Rembrandt 18
renewable energies 11
 resources 120
 systems 113
Rhowbotham, Kevin 14
RIBA Gold Medal 7
Rotterdam 26

Saltmarsh, Paul 105
Sassen, Saskia 10, 14
Scotland 55
Seckau 95
Sheffield 116, 117, 118
shopping 5, 6
Shoreditch 116, 117, 118
sick building syndrome 85
simulation 105, 106
Slot House 119, 120
Smith, M 87
solar energy 73, 101, 109
 radiation 77, 82
Spuybroek, Lars 18
Stalin viii
Stanners, D 87
Steadman 73

Index

Stockholm 89
Styria 95
suburb 24, 25, 26, 28, 34, 40, 41, 55, 89, 103, 118, 120
sulphur dioxide 84
surface emittance 104
sustainability, ecological 109
    economic 111
swarm models 18
Swiss Railways 47, 48
Switzerland 48

technoburbs 9, 24, 25
Texas 84
TGV 49
thermal capacity 104
topo-energetic modelling 107
Toulouse 75, 77
town planning 35, 46
transpiration 91
transport, public 11, 26, 46, 47, 49, 114
Trondheim 75
Trotsky viii

urban climate 81
    design 34
    ecology 107, 116
    footprint 3, 91
    intensification 119
    mines 61
    morphology 88, 91
    nature 117, 118
    transport planning 46
Urban Task Force 7, 24, 38

utopias 33

Vancouver 41
Vitruvius 102, 108

Walpole, Ken 39
Waring, George E 64, 68
waste recycling 39
web 15
Whitelegg, J 87
WHO 84
Wigley, Mark 94
Williams, N 87
wind power 78
Wright, Frank Lloyd 8

Zurich 46, 80
Zurich Urban Transport Authority 47